£190.00

7

..ES

D1759957

028484

HEAVY METALS RELEASE IN SOILS

HEAVY METALS RELEASE IN SOILS

Edited by

H. MAGDI SELIM
DONALD L. SPARKS

LEWIS PUBLISHERS

Boca Raton London New York Washington, D.C.

Library of Congress Cataloging-in-Publication Data

Heavy metals release in soils / edited by H. Magdi Selim, Donald L. Sparks.
   p. cm.
Includes bibliographical references and index.
ISBN 1-56670-531-2 (alk. paper)
   1. Soils—Heavy metal content. 2. Heavy metals—Environmental aspects. 3.
Soils Solute movement. I. Selim, Hussein Magdi Eldin, 1944- II. Sparks, Donald L.

S592.6.H43 H435 2001
631.4'1—dc21   2001029341
                                                                      CIP

Visit the CRC Press Web site at www.crcpress.com

© 2001 by CRC Press LLC
Lewis Publishers is an imprint of CRC Press LLC

No claim to original U.S. Government works
International Standard Book Number 1-56670-531-2
Library of Congress Card Number 2001029341
Printed in the United States of America   2 3 4 5 6 7 8 9 0
Printed on acid-free paper

Preface

Characterizing the nature of heavy metal release reactions, sorption mechanisms, and movement in the soil is the main topic of this book. Because soils are heterogeneous, heavy metals in soils can be involved in a series of complex chemical and biological interactions including oxidation-reduction, precipitation and dissolution, volatilization, and surface and solution phase complexation. The heterogeneous nature of the different soil constituents adds to the complexity of interactions of heavy metal species with the soil environment.

In the first four chapters, the primary focus is on transport processes and parameters which control the mobility of heavy metals in contaminated and uncontaminated soils, assessment of their potential for migration, and the impact on the soil environment. Models that are often used to describe the reactivity and transport of heavy metals in the soil system are described. Such models must account for all reaction and retardation of the contaminants relative to the groundwater. In addition to hydrologic parameters, such transport models require parameters that reflect the geochemical interactions of contaminants of concern with soils and aquifer materials. The reliability of predictions based on predictive models depends directly on the quality of the set of input parameters utilized. We recognize that comprehensive modeling of heavy metal sorption and release in soils is, therefore, a challenging task and should, ideally, take into account competitive sorption, dissolution and surface precipitation, sorbent heterogeneity, and the kinetics of these processes.

Chapter 1 provides a comprehensive treatment of multiple retention mechanisms of the kinetic and nonlinear type, which are frequently used to describe sorption-release of several heavy metals in soils. Hysteresis phenomena, equilibrium-kinetic reversible and irreversible reactions, ion exchange processes, and specific sorption of heavy metals are discussed. Model prediction capabilities of the slow release of copper are tested. In addition, behavior of vanadium, phosphorus, copper, and magnesium mobility from miscible displacement (transport) in several soils is described.

In Chapter 2, the fate and transport of heavy metals and radionuclides in zeolitized tuffs from the Nevada Test Site (NTS) are described. The ability of zeolitized tuffs to retard the migration of radionuclides such as strontium, cesium, and lead in transport columns is discussed. In addition, the behavior of cations and anions with different sorption affinities for different types of sorption sites are presented. The extent and implication of nonequilibrium behavior of radionuclide transport are also discussed.

In Chapter 3, various modeling approaches of heavy metal sorption in soil systems are presented, with emphasis on competitive sorption to pure soil components, such as clay minerals, oxides, and humic substances. Modeling approaches based on cation exchange and specific sorption are discussed. A case study on induced release of cadmium, zinc, and lead from smelter-contaminated soils is also presented. In Chapter 4, metal release and transport from soils contaminated by anthropogenic activities related to metal mining, extraction, and processing are presented. Various techniques and methodologies designed to evaluate the efficiency of chemical treatments for reducing metal release and transport from contaminated soils are also discussed. Chemical immobilization treatments are evaluated, using solute transport experiments with repacked soil columns. Agricultural limestone, mineral rock phosphate, and diammonium phosphate as chemical immobilization treatments are used to reduce Cd, Pb, and Zn solubility and transport in smelter-contaminated soil. The efficiency of the various immobilization treatments on heavy metal release is assessed, and mechanisms of metal release and the interactions between metal species in solution and the soil during transport processes are presented.

The subsequent two chapters of this book are devoted to the kinetics of sorption-release processes in the soil environment. Theoretical and experimental analyses of kinetic and reversible processes are presented. In Chapter 5, the technique of phase plane analysis is discussed and utilized to quantify the time for sorption processes to attain equilibria conditions. Kinetics of hysteresis in release or desorption are developed based on an elementary model using phase plane analysis. Furthermore, a qualitative approach is used to develop a new model for multilayer sorption on matrix surfaces. Kinetic aspects of operational methods of chemical extractions that characterize the stability of the various trace metal-soil constituent associations are presented in Chapter 6. An overview on possible artifacts of sequential extraction procedures is discussed. Differentiation of iron fractions, using kinetic results, of the labile and slowly labile forms is discussed.

Identification of the major soil parameters affecting metal lability in soils is requisite to prediction of metal behavior and establishment of appropriate soil screening levels. In Chapter 7, the focus is on major soil parameters and operational methods which influence the partitioning coefficient for heavy metals by different soils. Understanding how adsorption mechanisms are affected by reaction conditions is of considerable interest to developers of surface complexation models. In Chapter 8, mechanisms of sulfate adsorption on iron oxides and hydroxides are presented. The effects of pH, ionic strength, and reactant concentration on the formation of inner- and outer-sphere complexes over a wide range of surface loadings are discussed.

In Chapter 9, the focus is on sorption and release processes of selenate in various soils typical of the Mediterranean area. Factors influencing complex formation of arsenate are outlined. The role of pH and phosphate ions in arsenic remobilization in a historically polluted soil was emphasized. The extent of high affinity retention of selenate on sorption sites, with increasing selenate concentrations, is presented. Effect of concentrations of selenium on soil retention and selenium bioavailability in the environment is also presented. In subsequent chapters, complexation and speciation processes and their influence on heavy metal mobility are discussed in detail. Experiments to determine arsenic-bearing phases are discussed. Association with amorphous iron oxides and arsenate mobility based on sequential extractions are also described. Results of remobilization of arsenic based on batch experiments with or without pH control and column tests are also presented. Results of arsenic remobilization under alkaline and acid conditions are demonstrated.

We wish to sincerely thank the contributors of this book for their diligence and cooperation in achieving our goal and making this volume a reality. We are most grateful for their time and effort in critiquing the various chapters, and in maintaining the focus on the release of heavy metals in soils. Special thanks are due to Michael

Amacher, Nick Basta, Paul Bell, Feng Xiang Han, Dean Hesterberg, Steven McGowen, Lambis Papelis, Derek Peak, and Liuzong Zhou for their help in reviewing individual chapter contributions. Without the support of the Louisiana State University and the University of Delaware, this project could not have been achieved. We also express our thanks to Ms. Randi Gonzalez and the CRC/Lewis Publishers staff for their help and cooperation in the publication of this book.

H. Magdi Selim
Baton Rouge, Louisiana

Donald L. Sparks
Newark, Delaware
April 2001

The Editors

H. Magdi Selim is Professor of Soil Physics at the Louisiana State University Agricultural Center, Baton Rouge. He received his B.S. degree in Soil Science from Alexandria University in 1964, and his M.S. and Ph.D. in Soil Physics from Iowa State University in 1969 and 1971, respectively. Professor Selim has published numerous papers, reports, bulletins, and book chapters. He has authored one book and is the editor of several books and monographs. His research interests concern the modeling of the mobility of dissolved chemicals and their reactivity in soils and groundwater. He is the original developer of the two-site as well as the second-order multi-reaction models for the fate and transport and retention of heavy metals, radio-nuclides, explosive contaminants in soils, and natural materials in porous media. His research interests also include saturated and unsaturated water flow in multilay-ered soils. Professor Selim served as associate editor of *Water Resources Research* and the *Soil Science Society of America Journal.*

Dr. Selim is the recipient of several professional awards including the Phi Kappa Phi, the Gamma Sigma Delta Award for Research, the Joe Sedberry Graduate Teaching Award, the First Mississippi Research Award for Outstanding Achieve-ments in Louisiana Agriculture, and the Doyle Chambers Career Achievements Award. Professor Selim is a Fellow of the American Society of Agronomy and the Soil Science Society of America. He has served on several committees of the International Society of Trace Element Biogeochemistry, the Soil Science Society of America, and the American Society of Agronomy. He served as the Chair of the soil physics division of the Soil Science Society of America.

Donald L. Sparks is Distinguished Professor of Soil Science, Francis Alison Professor, and Chairperson, Department of Plant and Soil Sciences at the University of Delaware at Newark. He also holds joint faculty appointments in the Departments of Civil and Environmental Engineering and Chemistry and Biochemistry, and in the College of Marine Studies. He received his B.S. and M.S. degrees at the Uni-versity of Kentucky, Lexington, and his Ph.D. degree in 1979 from the Virginia Polytechnic Institute and State University, Blacksburg, VA. Dr. Sparks is interna-tionally recognized for his research in the areas of kinetics of soil chemical processes, surface chemistry of soils and soil components using *in situ* spectroscopic and microscopic techniques, and the physical chemistry of soil potassium. He has pio-neered the application of chemical kinetics to soils and soil minerals, including development of widely used methods, elucidation of rate-limiting steps and mech-anisms, and coupling of kinetic studies with molecular-scale investigations. He is the author or coauthor of 170 publications. These include 30 edited or authored books, 33 book chapters, and 105 refereed papers.

Dr. Sparks has received numerous awards and honors. He was named a Fellow of both the American Society of Agronomy and the Soil Science Society of America. He is a member of the American Society of Agronomy, Soil Science Society of America, Clay Minerals Society, American Association for the Advancement of Science, American Chemical Society, American Geochemical Society, and the hon-orary societies Gamma Sigma Delta and Sigma Xi. He has served on many com-mittees of both the Soil Science Society of America and the American Society of Agronomy and is Past President of the Soil Science Society of America.

Contributors

Herbert E. Allen
Department of Civil and Environmental
  Engineering
University of Delaware
Newark, DE

Michael C. Amacher
Rocky Mountain Station
USDA-Forest Service
Logan, UT

Nicholas T. Basta
Department of Plant and Soil Sciences
Oklahoma State University
Stillwater, OK

A. Bermond
Institut National Agronomique
  Paris–Grignon
Laboratoire de Chimie Analytique
Paris, France

Evert J. Elzinga
Department of Geosciences
State University of New York
  at Stony Brook
Stony Brook, NY

J.P. Ghestem
Institut National Agronomique
  Paris–Grignon
Laboratoire de Chimie Analytique
Paris, France

Feng Xiang Han
Department of Plant and Soil Science
Mississippi State University
Mississippi State, MS

Michael H.B. Hayes
Department of Chemical and
  Environmental Sciences
University of Limerick
Ireland

Isabelle Le Hécho
Université de Pau et des Pays de
  l'Adour–Hélioparc
France

Christopher A. Impellitteri
Department of Civil and Environmental
  Engineering
University of Delaware
Newark, DE

William L. Kingery
Department of Plant and Soil Science
Mississippi State University
Mississippi State, MS

Ruben Kretzschmar
Institute of Terrestrial Ecology
Swiss Federal Institute
  of Technology
Schlieren, Switzerland

Virginie Matera
Université de Pau et des Pays de
  l'Adour–Hélioparc
France

Steven L. McGowen
USDA-ARS
Waste Management and Forage
  Research Unit
Mississippi State, MS

Seth F. Oppenheimer
Department of Mathematics and
  Statistics
Mississippi State University
Mississippi State, MS

Charalambos Papelis
Desert Research Institute
University and Community College
  System of Nevada
Las Vegas, NV

Derek Peak
Department of Plant and Soil Sciences
University of Delaware
Newark, DE

G. Petruzzelli
Institute of Soil Chemistry
Pisa, Italy

B. Pezzarossa
Institute of Soil Chemistry
Pisa, Italy

Jennifer K. Saxe
Department of Civil and Environmental
  Engineering
University of Delaware
Newark, DE

H. Magdi Selim
Department of Agronomy
Louisiana State University
Baton Rouge, LA

André J. Simpson
Department of Chemistry
The Ohio State University
Columbus, OH

Donald L. Sparks
Department of Plant and
  Soil Sciences
University of Delaware
Newark, DE

Wooyong Um
Desert Research Institute
University and Community College
  System of Nevada
Las Vegas, NV

Andreas Voegelin
Institute of Terrestrial Ecology
Swiss Federal Institute
  of Technology
Schlieren, Switzerland

Yujun Yin
Crestron Electronics, Inc.
Rockleigh, NJ

Sun-Jae You
Department of Marine Environmental
  Engineering
Kunsan National University
Cheonbuk, Republic of Korea

Contents

CHAPTER 1

Sorption and Release of Heavy Metals in Soils: Nonlinear Kinetics

H. Magdi Selim and Michael C. Amacher

INTRODUCTION

Because soils are heterogeneous, numerous studies have focused on the interaction of several heavy metals with different soil constituents. Thus, an accurate description of the complex interactions of heavy metals in soils is a prerequisite to predicting their behavior in the vadose zone. Specifically, to predict the fate of heavy metals in soils, one must account for retention and release reactions of the various species in the soil environment. Heavy metals in soils can be involved in a series of complex chemical and biological interactions including oxidation-reduction, precipitation and dissolution, volatilization, and surface and solution phase complexation.

An example of a study that investigated the interaction of solutes with different soil constituents is that by Sequi and Aringhieri (1977), who reported that removal of organic matter released new sorption sites on soil. In contrast, Cavallaro and McBride (1984) found that treatment of clays by removal of organics tended to either enhance or have little effect on sorption and fixation of Cu and Zn. They attributed this behavior to dominance of oxides in the soil clays in the sorption and fixation of heavy metals compared to the organic component in soil. Atanassova (1995) investigated Cu adsorption-desorption in a vertisol and a planosol and their clay fractions and found that Cu sorption was well described by Freundlich and Langmuir isotherms. It was suggested that the observed decrease of the distribution coefficient (K_d) with increasing Cu concentration was due to a high affinity of the sorption sites for Cu at low surface coverage. Dependence of K_d on concentration is a manifestation of sorption heterogeneity and a clear indication of the nonlinearity of Cu sorption isotherms. Atanassova (1995) also showed that Cu desorption was nonhysteretic and fully reversible for the planosol, whereas strong desorption hysteresis was observed for the vertisol, which had a significantly higher organic matter

1-56670-531-2/01/$0.00+$1.50

content than the planosol. Wu et al. (1999) found that Cu was preferentially sorbed on organic matter associated with the coarse clay fraction. They also suggested that iron oxides block available sites by coating lateral surfaces of layer silicates. Wu et al. (1999) also suggested that observed adsorption-desorption hysteresis is probably due to extremely high energy bonding with organic matter and layer silicate surfaces.

Heavy metal isotherms for different soils as well as clay fractions often exhibit a nonlinear shape. Moreover, most investigations do not consider the kinetic behavior of heavy metal retention. Since the nonlinearity of heavy metal retention directly influences their mobility in soils (Selim and Amacher, 1997), knowledge of the time-dependence nature of their behavior is important. It is equally important that reactions that control the release of heavy metals already adsorbed or retained by the soil be quantified. Observations from kinetic and transport experiments often indicate a slow and continued heavy metals release at low concentrations for extended periods of time. Such slow release or excessive tailing of breakthrough curves is often difficult to describe and single reaction type models fail to quantify the results. An example of such a slow release is that of drainage water from abandoned mines, which often contain high Cu and other heavy metal concentrations and can be of environmental concern.

In this chapter we present multireaction and competitive retention models consisting of nonlinear and kinetic reactions. Multireaction models are empirical and include linear and nonlinear equilibrium and reversible and irreversible kinetic retention reactions. We also present a multicomponent model in which competitive equilibrium or kinetic ion exchange is the primary sorption-desorption mechanism, but an irreversible (sink/source) reaction is also included to describe a specific sorption process. Limitations of these approaches and case studies for sorption and release of several heavy metals are presented. Examples of breakthrough curves for a copper (Cu)–magnesium (Mg) system and a vanadium (V)–phosphorus (P) system are also discussed.

MULTIREACTION KINETIC MODEL

Model Formulation

The multireaction approach, often referred to as the multisite model, acknowledges that the soil solid phase is made up of different constituents (clay minerals, organic matter, iron, and aluminum oxides). Moreover, a heavy metal species is likely to react with various constituents (sites) via different mechanisms (Amacher et al., 1988). As reported by Hinz et al. (1994), heavy metals are assumed to react at different rates with different sites on matrix surfaces. Therefore, a multireaction kinetic approach is used to describe heavy metal retention kinetics in soils. The multireaction model used here considers several interactions of one reactive solute species with soil matrix surfaces. Specifically, the model assumes that a fraction of the total sites reacts rapidly or instantaneously with solute in the soil solution, whereas the remaining fraction reacts more slowly with the solute. As shown in Figure 1.1, the model includes reversible as well as irreversible retention reactions that occur concurrently and consecutively. We assume a heavy metal species is

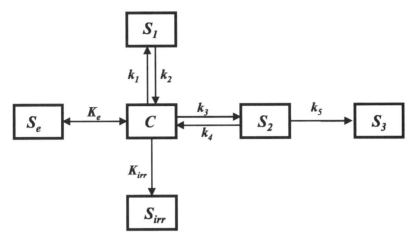

Figure 1.1 Schematic of the multireaction model for heavy metals reactivity in soils.

present in the soil solution phase, C (mg L^{-1}), and in several phases representing metal species retained by the soil matrix designated as S_e, S_1, S_2, S_s, and S_{irr} (mg kg^{-1} soil). We further considered the sorbed phases S_e, S_1, and S_2 are in direct contact with the solution phase (C) and are governed by concurrent reactions. Specifically, C is assumed to react rapidly and reversibly with the equilibrium phase (S_e) such that

$$S_e = K_e \left(\frac{\Theta}{\rho} \right) C^n \qquad (1.1)$$

where k_e is a distribution coefficient (m^3 Mg^{-1}), ρ is soil bulk density (Mg/m^3), Θ is water content (m^3/m^3), and n is the reaction order (dimensionless). Moreover, n represents a nonlinearity parameter which is commonly less than unity (Buchter et al., 1989). This parameter represents the heterogeneity of sorption sites having different affinities for heavy metal retention on matrix surfaces (Kinniburgh, 1986). The relations between C and the sorbed phases S_1 and S_2 were assumed to be governed by nonlinear kinetic reactions expressed as

$$\frac{\partial S_1}{\partial t} = k_1 \frac{\Theta}{\rho} C^m - k_2 S_1 \qquad (1.2)$$

$$\frac{\partial S_2}{\partial t} = \left[k_3 \frac{\Theta}{\rho} C^m - k_4 S_2 \right] - k_5 S_2 \qquad (1.3)$$

where t is time (h), k_1 and k_2 are the rate coefficients (h^{-1}) associated with S_1, and m is the reaction order. Similarly, for the reversible reaction between C and S_2, k_3 and k_4 are the respective rate coefficients (h^{-1}). In the above equations, we assume n = m, since there is no known method for estimating n and/or m independently.

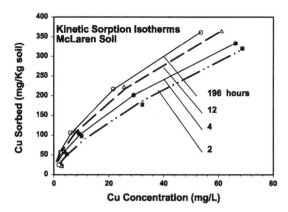

Figure 1.2   Adsorption isotherms for Cu on McLaren soil for different retention times. The solid curves are based on the Freundlich equation.

The multireaction model accounts for irreversible retention in two ways. First, as a sink term Q, which represents a direct reaction between the solution phase C and S_{irr} (e.g., precipitation/dissolution or immobilization) as a first-order kinetic process, where k_{irr} is the associated rate coefficient (h^{-1}).

$$Q = \rho \, \frac{\partial S_{irr}}{\partial t} = k_{irr} \Theta C \tag{1.4}$$

Irreversible retention was also considered to be the result of a subsequent reaction of the S_2 phase into a less accessible or strongly retained phase S_3 such that,

$$\frac{\partial S_3}{\partial t} = k_5 S_2 \tag{1.5}$$

One may regard the slowly reversible phase S_3 as a consequence of rearrangement of solute retained by the soil matrix. Mechanisms associated with irreversible reactions include different types of surface precipitation, which accounts for the formation or sorption of metal polymers on the surface, a solid solution or coprecipitate that involves co-ions dissolved from the sorbent, and a homogeneous precipitate formed on the surface composed of ions from the bulk solution or their hydrolysis products (Farley et al. 1985). The continuum between surface precipitation and chemisorption is controlled by several factors including: (1) the ratio of the number of sites to the number of metal ions in solution; (2) the strength of the metal-oxide bond; and (3) the degree to which the bulk solution is undersaturated with respect to the metal hydroxide precipitate. Such mechanisms are consistent with one or more irreversible reactions associated with our model presented in Figure 1.1.

An example of the kinetic behavior of heavy metal sorption by soils is shown in Figure 1.2. Here, several sorption isotherms representing the total amount of Cu retained (S) vs. Cu in soil solution (C) are shown for the McLaren soil (coarse-loamy over sandy or sandy-skeletal, mixed, super active, Typic Eutrocryepts, pH

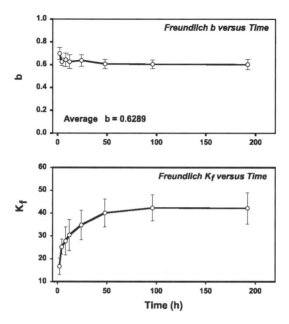

Figure 1.3   Freundlich parameters K_f and b vs. retention time for Cu retention in a McLaren soil.

4.1, 3.03% organic matter content, and cation exchange capacity [CEC] of 33.0 $mmol_c$ Kg^{-1}). McLaren soil was obtained from a site near an abandoned Cu mine on Fisher Mountain, Montana. Acid mine drainage flows into Daisy Creek below the mine, which is located about 2 miles from Cooke City, Montana (Amacher et al. 1995). Results from an undisturbed soil located near the minespoil are reported here. These results were obtained for selected reaction times from batch experiments as described by Amacher et al. (1988). Initial concentrations (C_o) of Cu were 5, 10, 20, 50, and 100 mg L^{-1} and the sorption reaction times were 2, 4, 8, 12, 20, 48, 96, and 192 h. Reagent-grade copper perchlorate was used, and the background solution composition was 0.005 M $Mg(ClO_4)_2$. The results in Figure 1.2 show the extent of Cu retention kinetics in this soil. In addition, Cu retention is not only time-dependent but also nonlinear. Similar kinetic and nonlinear sorption behavior of other heavy metals has been observed by numerous investigators. The concave sorption nonlinearity implies that Cu mobility in the soil solution tends to increase as the concentration increases. A simple way to describe these sorption isotherms is by use of the Freundlich equation:

$$S = K_f C^b \tag{1.6}$$

where K_f is the distribution coefficient (cm^3 g^{-1}) and b is an exponential parameter. A set of K_f and b values was obtained for each S vs. C data set. Examples of isotherms calculated using Equation 1.6 are shown by the solid curves in Figure 1.2. The values obtained indicate that K_f was strongly time dependent (Figure 1.3). The K_f values along with their standard errors are also given in Table 1.1. The goodness

Table 1.1 **Parameter Estimated, Standard Errors, and Coefficient of Correlation (r^2)**
for the Langmuir and Freundlich Equilibrium Models vs. Retention Time

Time (h)	S_{max} (mg/kg)	ω	r^2	K_f (L/mg)	b	r^2
2	637.8 ± 163.7	71.2 ± 30.9	0.976	16.71 ± 3.51	0.696 ± 0.053	0.988
4	554.2 ± 71.6	42.4 ± 10.9	0.986	25.25 ± 3.32	0.624 ± 0.634	0.994
8	634.9 ± 83.9	41.1 ± 10.7	0.987	27.82 ± 6.15	0.640 ± 0.058	0.983
12	619.6 ± 75.9	36.7 ± 9.2	0.987	30.41 ± 6.76	0.625 ± 0.058	0.981
24	721.9 ± 85.9	36.2 ± 8.5	0.989	34.79 ± 6.44	0.636 ± 0.050	0.987
48	683.2 ± 62.9	30.0 ± 5.9	0.992	40.09 ± 6.10	0.606 ± 0.041	0.993
96	701.8 ± 78.3	29.3 ± 6.96	0.988	42.31 ± 5.75	0.603 ± 0.037	0.994
196	689.68 ± 82.70	28.5 ± 7.40	0.986	42.16 ± 6.86	0.601 ± 0.045	0.989

of fit of the model to the data is given by the r^2 values. The K_f value increased from 16.7 to 42.1 cm^3 g^{-1} as retention time increased from 2 and 196 h, respectively. The parameter b was less than unity for all isotherms and ranged from 0.601 to 0.696 with a mean value of 0.629. No consistent trend in b with reaction time was observed. Similar results were observed for Cd by Hinz and Selim (1994), who found that b varied from 0.60 to 0.74 for a time range of 2 to 240 h.

Nonlinearity and Kinetics

The extent of nonlinearity of Cu isotherms can be compared to Cu isotherms obtained by other investigators. For example, Buchter et al. (1989) measured K_f and b for Cu after 1 d of retention for 11 soils having a wide range of properties. They reported a range of b values from 0.47 to 1.42 with a mean value of 0.76. Recently, a somewhat lower value for b of 0.42 was reported by Houng and Lee (1998). These investigators also reported that b for Cu was not affected by the presence of Cd as a competing heavy metal. It should be emphasized that in most studies, sorption isotherms represent 1 d of reaction between solute and soil, and the kinetics of the sorption processes are not usually investigated.

The results shown in Figure 1.3 indicate that for a given reaction time, the Freundlich equation (1.6) was capable of describing the overall shape of the isotherms. Nevertheless, the time dependence of K_f implies that the model given by Equation 1.6 represents an oversimplification of the retention mechanisms in the soil. Specifically, a single equilibrium reaction was incapable of describing the kinetic behavior shown.

Measured isotherms of the type shown in Figure 1.2 are also frequently described using the Langmuir approach. Several investigators indicated that 1 d isotherms for Cu and other heavy metals can be well described using both Freundlich and Langmuir equations (Cavallaro and McBride, 1978; Harter, 1984; Amacher et al., 1986; Schulte and Beese, 1994; Atanassova, 1995; Houng and Lee, 1998). Although the Langmuir form is not strictly recommended over the Freundlich equation, it is often preferred since it provides a sorption maxima S_{max}, which can be correlated to intrinsic soil properties such as mineralogical composition, specific surface area, etc. Therefore, the isotherms shown in Figure 1.2 were also described using the Langmuir form,

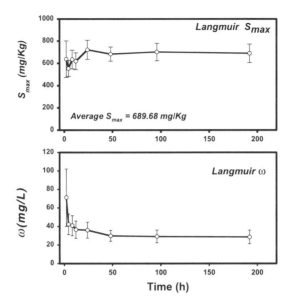

Figure 1.4   Langmuir parameters S_{max} and ω vs. retention time for Cu retention in a McLaren soil.

$$S = \frac{S_{max}C}{\omega + C} \qquad (1.7)$$

where the parameter ω (mg L^{-1}) is a Langmuir coefficient which represents a measure of the affinity of sites or bonding strength. Nonlinear, least-squares, best-fit Langmuir parameters for the different sorption times were calculated and are given in Table 1.1, and ω and S_{max} vs. retention time are shown in Figure 1.4. Based on r^2 and root mean square errors (rmse), the Langmuir formulation is capable of describing our Cu sorption data. The Langmuir approach was used successfully to describe sorption isotherms for several heavy metals including Cu (Cavallaro and McBride, 1978; Harter, 1984; Amacher et al. 1986; Atanassova, 1995; Houng and Lee, 1998).

Unlike the isotherms presented here, the time of reaction for most Cu isotherms reported in the literature is commonly limited to one day. The fact that the sorption capacity term (S_{max}) given in Table 1.1 did not appreciably change with retention time gives credence to the Langmuir model in describing our Cu results. This is a significant finding if sorption capacity is to be used to describe retention behavior as well as potential mobility of a heavy metal such as Cu in soils. If S_{max} is not a unique value and varies with time or concentration, the use of such a term in modeling efforts becomes questionable. We recognize that based on parameter estimation alone, we cannot conclude that the Langmuir equation is uniquely applicable to our Cu sorption data. As argued by Sposito (1982) and Amacher et al. (1988), isotherm equations such as Equations 1.6 and 1.7 should be regarded as empirical equations, and their use constitutes primarily a curve-fitting procedure. Based on goodness of fit of the sorption data, both the Freundlich and Langmuir approaches described the data equally well for all reaction times (Table 1.1). Such a finding is expected since both approaches result from the general isotherm equation (Kinniburgh, 1986):

$$\frac{S}{S_{max}} = \sum_{i=1}^{j} F_i \left[\frac{\left(\kappa_i C_i\right)^{\lambda_i}}{1 + \varepsilon_i \left(\kappa_i C_i\right)^{\lambda_i}} \right]^{\beta_i} \qquad (1.8)$$

where $j = 1$, $F_i = 1$, $\beta_i = 1$, $\lambda_i = 1$, and $\varepsilon_i = 1$ for the Langmuir equation. For the Freundlich equation, $j = 1$, $F_i = 1$, $\beta_i = 1$, and $\varepsilon_i = 0$.

Results from the kinetic batch experiments for McLaren soil are presented to show the changes in Cu concentration (C) vs. time for the various initial (input) concentrations (C_O). Retention of Cu by the soil matrix was rapid during the initial stages of reaction and was then followed by slow and continued Cu retention as depicted by the changes of C vs. time results. The capability of the multireaction approach to describe the experimental batch data is shown by the solid curves of Figure 1.5 for the various initial concentrations (C_o's). Good model predictions were observed for the wide range of input concentration values considered.

The multireaction model accounts for several interactions of the reactive solute species (Cu) within the soil system. Specifically, the model assumes that a fraction of the total sites reacts instantaneously with solute in the soil solution, whereas the remaining fraction is highly kinetic. As illustrated in Figure 1.1, the model also accounts for concurrent or consecutive irreversible solute sorption by the soil. As a result, different versions of the multireaction model shown in Figure 1.1 represent different reactions that can be used to describe Cu retention mechanisms. Several variations were examined: (1) a two-parameter model with k_e and k_{irr}, (2) a four-parameter model with k_e, k_1 and k_2, and k_{irr}, (3) another four-parameter model with k_e, k_3, k_4, and k_5, and (4) a five-parameter model with k_1, k_2, k_3, k_4, and k_{irr}. These variations were chosen based on our observations of C vs. time and assume the presence of at least one fraction of retention sites that reacts slowly (kinetic) and another that is kinetic but irreversible or slowly reversible in nature. Each model variation was fitted to the experimental data using a nonlinear, least-squares, parameter optimization scheme (van Genuchten, 1981). Criteria used for estimating the goodness-of-fit of the model to the data were the r^2 and the root mean square error (rmse) statistics (Kinniburgh, 1986),

$$rmse = \left[\frac{rss}{N - P} \right]^{1/2} \qquad (1.9)$$

where rss is the residual sum of squares, N is the number of data points, and P is the number of parameters.

Generally, individually fitted parameters had large standard errors due to small degrees of freedom. Several model variations even failed to fit Cu concentration vs. time data for most C_o's, and convergence was often not achieved. In addition, large variations existed among fitted model parameters. The example shown in Table 1.2 is for the simplest case where only two sorbed phases were considered: a reversible

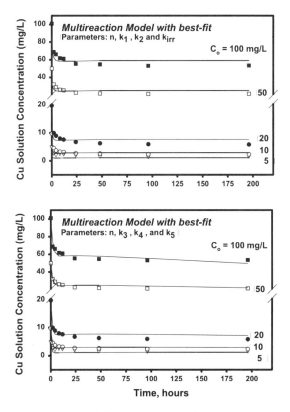

Figure 1.5   Experimental results of Cu in soil solution for McLaren soil vs. time for a wide range of initial concentrations (C_o's). The solid curves were obtained using the multireaction model with concurrent (top figure) and consecutive (bottom figure) irreversible reactions. Fitted parameters were n, k_1, k_2, and k_{irr} (top) and n, k_3, k_4, and k_5 (bottom).

Table 1.2   Comparison of Goodness of Fit of the Multireaction Model with S_e and S_{irr} for McLaren Soil. (All C_o's were used for the 'overall' fit.)

C_o mg/L	Model r^2	RMSE	n	SE	K_e (m^{-3}/Mg)	SE	K_{irr} (h^{-1})	SE
5	0.747	2.024	0.244	831.6	0.010	19.28	0.0138	0.0535
10	0.641	5.460	0.594	15,338	0.002	98.49	0.0113	0.1033
20	0.757	9.349	0.221	4200	0.002	39.46	0.147	0.0277
50	0.501	23.034	0.212	11,609	0.006	394.9	0.0043	0.0305
100	0.550	38.126	0.295	9265	0.004	273.8	0.0035	0.0077
Overall-A	0.981	3.098	0.584	0.0655	3.605	0.917	0.0028	0.0004
Overall-B	0.981	3.075	0.629[a]	—	3.041	0.135	0.0028	0.0004

[a] n = b was obtained from Table 1.1 where an average b was used.

Table 1.3 Goodness of Fit of the Multireaction Model when the Nonlinear Parameter n Was Derived from the Freundlich Equation (n = b = 0.629) for Individual as Well as Overall C_o's

C_o mg/L	Model r^2	RMSE	k_I	SE	K_2	SE	K_{irr}	S_E
					h^{-1}			
6	0.601	0.3601	0.682	0.225	0.429	0.167	0.0062	0.0022
10	0.849	0.3668	0.922	0.109	0.270	0.043	0.0079	0.0022
20	0.762	0.8887	1.133	0.179	0.315	0.063	0.0059	0.0018
50	0.802	2.1709	0.994	0.163	0.273	0.056	0.0028	0.0010
100	0.663	4.3336	1.217	0.252	0.357	0.087	0.0019	0.0007
Overall	0.992	2.0635	1.131	0.089	0.323	0.031	0.0020	0.0003

(S_e) and an irreversible (S_{irr}) where three model parameters were estimated (n, k_e, and k_{irr}). For most C_o's, parameter optimization failed to fit the data. Moreover, the nonlinear parameter n was particularly difficult to estimate due to extremely high standard errors. We also found that when other versions of the multireaction model were tested, poor parameter estimates were obtained. Therefore, the use of a data set from a single initial input concentration is not recommended in this study for the purpose of parameter estimation. To overcome this difficulty, the entire data set consisting of all input concentrations (C_o's) was used in the nonlinear least-square optimization. The resulting overall set of parameter estimates for our simplest model version (e.g., n, k_e, and k_{irr}) is given in Table 1.2. This use of the entire data set increased the degrees of freedom. This decreased the root mean square error (rmse) and improved r^2 value as well as decreased parameter standard errors. In addition, the overall shape of C vs. time curves for the different C_o's were improved using this overall fitting strategy. The estimated value for n was 0.584, which is within the confidence interval of the Freundlich b (0.629) obtained earlier (Figure 1.2 and Table 1.1). Consequently, we tested whether the value of Freundlich b can be used in place of the parameter n in the multireaction model. For this model version, only two parameters were estimated (k_e and k_{irr}), which resulted in an equally good description of C vs. time for all C_o's (figure not shown). Moreover, when n = b was assumed, the resulting rmse and r^2 values were comparable to those when n was estimated (Table 1.2).

Based on the above findings, we tested other versions of the multireaction model where n = b was assumed. Based on goodness of fit of the model to the experimental results, the use of Freundlich b in place of n is recommended only if the entire data set (for all C_o's) is simultaneously used in the optimization method (Table 1.3). Obviously, the use of n = b reduces the number of total parameters to be estimated, which is advantageous because the parameter n is, in general, difficult to estimate, as discussed above. It is obvious from the simulations shown in Figure 1.5 that a number of model versions were capable of producing indistinguishable simulations of the data. Similar conclusions were made by Amacher et al. (1988, 1990) for Cd, Cr(VI), and Hg retention by several soils. They also stated that it was not possible to determine whether the irreversible reaction is concurrent or consecutive, since both model versions provided a similar fit of their batch data. Contrary to their findings, we found that the use of a consecutive irreversible reaction provided an improved fit of our Cu data over other model versions. This finding is based not only on r^2 and rmse values but also on visual observation of model simulations.

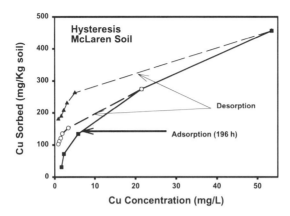

Figure 1.6 Adsorption and desorption or release isotherms for McLaren soil.

Release

Desorption or release results for McLaren soil are presented along with sorption isotherms in the traditional manner in Figure 1.6 and clearly indicate considerable hysteresis for Cu sorption/desorption. Release or desorption was carried out using a successive dilution technique (Amacher et al., 1988). Following the last sorption step, the soil residue with initial Cu concentrations (C_O) of 50 and 100 mg L^{-1} was extracted with 0.005 M $Mg(ClO_4)_2$ several times. Each extraction was conducted by adding 30 ml of 0.005 M $Mg(ClO_4)_2$ solution after decanting the supernatant. The first four extractions were completed in 2 hours to evaluate the equilibrium adsorption sites. The subsequent four extractions were conducted with a 3-day equilibration time interval between each desorption step to assess the affinity of retention sites for Cu during desorption.

As indicated by Figure 1.6, the hysteretic behavior resulting from discrepancy between sorption and desorption isotherms was not surprising in view of the kinetic retention behavior of Cu in this McLaren soil. Several studies have shown that observed hysteresis in batch experiments may be due to kinetic retention behavior, slow release, and/or irreversible sorption. Sorption-desorption isotherms indicate that the amount of irreversible or nondesorbable phases increased with time of reaction. Copper may be retained by heterogeneous sites with a wide range of binding energies. At low concentrations, binding may be irreversible. The irreversible amount almost always increased with time (Harter, 1984). Wu et al. (1999) suggested that Cu hysteresis is probably due to extremely high energy bonding with organic matter and layer silicate surfaces. The fraction of nondesorbable Cu was referred to as specifically sorbed. Atanassova (1995) showed that desorption was nonhysteretic for a planosol, whereas strong desorption hysteresis was observed for a vertisol. It was suggested that Cu was fixed in a nonexchangeable form, which resulted in lack of reversibility as well as hysteretic behavior.

Several studies reported that the magnitude of sorption/desorption hysteresis increases with longer sorption periods. Ainsworth et al. (1994) found that despite increasing desorption times from 16 hours to 9 weeks, sorption/desorption hysteresis

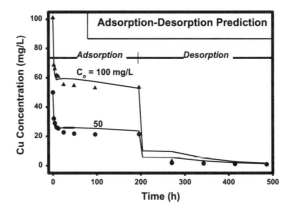

Figure 1.7 Experimental results of Cu in soil solution for McLaren soil vs. time during adsorption and desorption or release for initial concentration C_o = 50 and 100 mg L^{-1}.
The solid curves are predictions based on parameters from adsorption data using
the multireaction model. The irreversible mechanism was accounted for based on
concurrent reactions.

persisted for Co and Cd on hydrous ferric oxide. They also found that Cd and Co
sorption displayed increasing hysteretic behavior upon aging from 2 to 16 weeks,
while Pb sorption was reversible. Hysteresis has also been observed in exchange
reactions for several cations, where the exchange of one sorbed cation with another
is not completely reversible, i.e., the forward and reverse exchange reactions do not
result in the same isotherms. The hysteretic behavior of cation exchange is abun-
dantly reported in the literature. A critical review was published by Verburg and
Baveye (1994). From a survey of the literature, they were able to categorize several
elements into three categories. The elements in each category were found to show
hysteretic exchange between groups, but not within groups. Verburg and Baveye
(1994) proposed that exchange reactions are most likely a multistage kinetic process
in which the later rate-limiting processes are a result of physical transformation in
the system, e.g., surface heterogeneity, swelling hysteresis, and formation of quasi-
crystals, rather than simply a slow kinetic exchange process where there exists a
unique thermodynamic relationship for forward and reverse reactions.

Copper release results for McLaren soil were predicted using the multireaction
kinetic model and are illustrated by the solid curves shown in Figure 1.7. The data
are presented as C vs. time for two initial concentrations (C_o) of 50 and 100 mg/L.
Here all model parameters were based on the sorption data alone. Based on these
predictions, we can conclude that the multireaction model predicted Cu desorption
or release behavior satisfactorily. Predictions of desorption isotherms were not considered adequate at the initial stages of desorption. In addition, the model underpre-
dicted amounts sorbed, which directly influences subsequent predictions of the desorption isotherms. Discrepancies between experimental and predicted values are expected
if the amounts of Cu in the various phases (C, S_e, S_1, and S_2) at each desorption step
were significantly different. These underpredictions also may be due to the inherent
assumptions of the model. Nevertheless, the multireaction model provided adequate
predictions of Cu concentrations especially at large times of release.

MULTICOMPONENT MODEL

Model Formulation

We present a multicomponent transport model in which two principal reaction mechanisms are considered as the dominant metal retention processes in the soil system: (1) ions are adsorbed as readily exchangeable ions, and (2) ions are sorbed with high affinity to specific sites on the soil matrix. The first type of reaction is generally referred to as ion exchange, whereas the second type of reaction includes processes that cause strong heavy metal retention in the soil system. Such reactions may include the formation of inner-sphere complexes, surface precipitation, and possibly the penetration of heavy metal cations into the lattices of soil minerals. For convenience, we refer to the second type of reactions as specific sorption in a similar fashion to the terminology used by Tiller et al. (1984a), Selim et al. (1992), and Sparks (1999).

Ion exchange may be considered a fast or instantaneous nonspecific sorption mechanism and is a fully reversible reaction between heavy metal ions in the soil solution and those retained on charged surfaces of the soil matrix. In contrast, specific sorption is considered a second-order kinetic process involving metals with high affinity for the solid phase. Because of such high affinity, retention via specific sorption is assumed to be an irreversible or weakly reversible process. For a multi-ion system, incorporation of the two retention mechanisms, ion exchange and specific sorption, into the convective-dispersion transport equation for a heavy metal ion i yields,

$$\Theta \frac{\partial C_i}{\partial t} + \rho \left(\frac{\partial \Gamma_i}{\partial t} + \frac{\partial \Psi_i}{\partial t} \right) = \Theta D \frac{\partial^2 C_i}{\partial z^2} - v \frac{\partial C_i}{\partial z} \tag{1.10}$$

where Γ_k and Ψ_k are the amount retained on the exchanger surfaces and that specifically sorbed (mmol$_c$ kg^{-1}), respectively, and C_i is the concentration in solution (mmol$_c$ L^{-1}). We first describe the rate of retention term ($\partial \Gamma_i / \partial t$), which is governed by ion exchange. In a standard mass action formulation, the exchange reaction for two competing ions i and j may be written as (Selim et al., 1992),

$$^T K_{ij} = \frac{\left(a_i^* / a_i \right)^{vj}}{\left(a_j^* / a_j \right)^{vi}} \tag{1.11}$$

where $^T K_{ij}$ denotes the thermodynamic equilibrium constant, v (omitting the subscripts) is the valency, and a and a* are the ion activity in soil solution and on the exchanger surfaces, respectively. Based on equation [1.11], one can denote the parameter $^v K_{ij}$ as

$$^v K_{ij} = \frac{^T K_{ij}}{\left(\dfrac{(\zeta_i)^{vj}}{(\zeta_j)^{vi}} \right)} \tag{1.12}$$

where ^vK is the Vanselow selectivity coefficient and ζ the activity coefficient on the exchanger surfaces. If one restricts the analysis to binary homovalent exchange (e.g., Cu and Mg ions), then $v_i = v_j = v$. Furthermore, assuming similar ion activities in the solution phase, rearrangement of Equation 1.12 yields

$$K_{ij} = \left[^vK_{ij} \right]^{1/v} = \frac{\left(X_i / Y_i \right)}{\left(X_j / Y_j \right)} \tag{1.13}$$

where K_{ij} is a generic selectivity coefficient of ions i over j (Rubin and James, 1973) or a separation factor for the affinity of ions on exchange sites. In Equation 1.13, X_i is the equivalent fraction of ion i on the exchanger phase,

$$X_i = \Gamma_i / \Omega \tag{1.14}$$

where Γ_i is the amount of ion i retained on the exchanger surfaces and is expressed in $mmol_c$ kg^{-1}, Ω is the cation exchange (or sorption) capacity (CEC) of the soil ($mmol_c$ kg^{-1}), and Y_i is the equivalent ionic fraction in soil solution:

$$Y_i = C_i / C_T \tag{1.15}$$

where $C_T = \Sigma\ C_i$ is the total concentration ($mmol_c$ L^{-1}) and is the sum of C_1 and C_2 ion species for our binary system. Rearrangement of Equation 1.13 and assuming CEC to be time invariant yields the following isotherm relation for the amount sorbed as an equivalent fraction Γ vs. solution concentration C for ion i,

$$X_i = \frac{K_{12}Y_i}{1 + \left(K_{12} - 1 \right)Y_i} \tag{1.16}$$

For K_{ij} different from 1, Equation 1.16 indicates nonlinear ion exchange. Upon differentiation of Equation 1.16, the retention term ($\partial\Gamma_i/\partial t$) of Equation 1.10 for binary homovalent ions, at a constant total solution concentration C_T, is

$$\frac{\partial\Gamma_i}{\partial t} = \Omega\ \frac{\partial X_i}{\partial t} = \frac{K_{12}\Omega C_T}{C_T + \left(K_{12} - 1 \right)^2 C_i}\ \frac{\partial C_i}{\partial t} \tag{1.17}$$

and can be solved simultaneously with the convection-dispersion equation to obtain a solution for two ions in the soil system.

Equilibrium Ion Exchange Isotherms

Examples of ion exchange isotherms are illustrated in Figures 1.8 and 1.9 for a Cu-Mg system for two soils, Cecil (clayey, kaolinitic, thermic Typic Hapludult) and

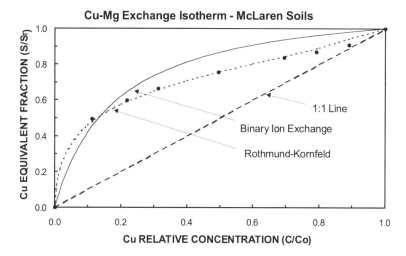

Figure 1.8   Copper-magnesium exchange isotherm for McLaren soil. Solid and dashed curves are simulations based on best-fit selectivity coefficient K_{12} of Equation 1.16 and that of the Rothmund–Kornfeld Equation 1.18.

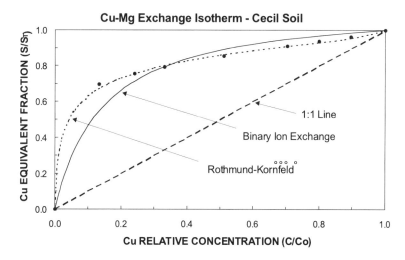

Figure 1.9   Copper-magnesium exchange isotherm for Cecil soil. Solid and dashed curves are simulations based on best-fit selectivity coefficient K_{12} of Equation 1.16 and that of Rothmund–Kornfeld Equation 1.18.

McLaren. The isotherms are represented as relative concentration (C/C_T) for Cu in the soil solution vs. the amount of Cu sorbed expressed as a fraction on the exchanger surface (S/S_T). Here C_T and S_T (=CEC) represent the total concentration in solution and that on the exchange surfaces. The methodology used to obtain the results shown was described by Selim et al. (1992) and Selim and Amacher (1997), where each soil was equilibrated with Cu and Mg at varying ratios. The samples were shaken for 24 h on a reciprocal shaker with various proportions of $Cu(ClO_4)_2$ and $Mg(ClO_4)_2$ solutions. The solutions were then centrifuged and decanted. For the first two steps (24 h each step), total concentration was $0.5\ M$ followed by four time steps at $0.005\ M$. Adsorbed cations were removed by three extractions with $1\ M$ NaOAc, and corrections were made for the entrained solution. Copper and Mg in the equilibrating and extraction solutions were analyzed by inductively coupled plasma (ICP) spectroscopy.

Measured results of Cu-Mg exchange isotherms for the two soils are shown in Figures 1.8 and 1.9. Continuous curves were calculated using the isotherm Equation 1.16 for homovalent binary ions. For both soils, sorption of Cu vs. Mg is preferred, and the isotherms are convex ($K_{CuMg} > 1$), indicating a high affinity to the exchange sites. The diagonal line represents a nonpreference isotherm ($K_{CuMg} = 1$) where competing ions (Cu-Mg) have equal affinity for exchange sites. For Cecil soil, a selectivity coefficient K_{CuMg} of 7.857 ($r^2 = 0.965$) was calculated, whereas a value of K_{CuMg} of 6.519 ($r^2 = 0.956$) was obtained for McLaren soil. The similarity of these values indicates that both soils have a similar affinity for Cu in the presence of Mg as the counter ion. For a Cd-Ca system, Selim et al. (1992) reported a value of K_{CdCa} of 2 for a Eustis sandy soil.

Equilibrium ion exchange reactions, similar to those discussed above, have been employed to describe sorption of heavy metals in soils by several investigators (Abd-Elfattah and Wada, 1981; Harmsen, 1977; Bittel and Miller, 1974; Selim et al., 1992; Hinz and Selim, 1994). For several heavy metals, the affinity for exchange sites increased with decreasing heavy metal fraction on exchanger surfaces. Using an empirical selectivity coefficient, it was shown that Zn affinity increased up to two orders of magnitude for low Zn surface coverage in a Ca background solution (Abd-Elfattah and Wada, 1981).

The Rothmund–Kornfeld approach incorporates variable selectivity based on the amount adsorbed (s_i) or exchanger composition. The approach is empirical and provides a simple equation that incorporated the characteristic shape of binary exchange isotherms as a function of s_i as well as the total solution concentration in solution (C_T). Harmsen (1977) and Bond and Phillips (1990) expressed the Rothmund–Kornfeld as

$$\frac{s_i}{s_T} = \frac{\kappa (c_i/c_j)^\sigma}{1 + \kappa (c_i/c_j)^\sigma} \tag{1.18}$$

where σ is a dimensionless empirical parameter associated with the ion pair i-j, s_T is the total amount sorbed (CEC), and κ (or $^R K_{ij}$) is the Rothmund–Kornfeld selectivity coefficient. The above equation is best known as a simple form of the Freundlich equation applied to ion exchange processes. As pointed out by Harmsen (1977), the Freundlich equation may be considered as an approximation of the Rothmund–Kornfeld equation valid for $s_i \ll s_j$ and $c_i \ll c_j$, where

$$s_i = \kappa (c_i)^\sigma \tag{1.19}$$

The Cu sorption isotherms in Figures 1.8 and 1.9 show the relative amount of Cu adsorbed as a function of relative solution concentration along with best-fit isotherms based on the Rothmund–Kornfeld Equation 1.18. For Cecil soil, best-fit values for κ and σ were 6.077 and 0.5775, respectively. For McLaren soil, the respective values were 3.127 and 0.5612. For both Cecil and McLaren soil, the use of a simple selectivity relation (Equation 1.12) was inadequate, whereas the Rothmund–Kornfeld proved superior for the entire concentration range. A similar approach was used by Hinz and Selim (1994) for Zn-Ca and Cd-Ca isotherms for two acidic soils. Their results showed sigmoidal shapes of the isotherms, which revealed that Cd as well as Zn sorption exhibited high affinity at low concentrations, whereas Ca exhibited high affinity at high heavy metal concentrations. A similar finding can be observed, to some extent, for Cu in McLaren soil (Figure 1.9).

Kinetic Ion Exchange

Several studies have indicated that ion exchange is a kinetic process in which equilibrium was not reached instantaneously. Sparks (1989) compiled an extensive list of cations (and anions) that exhibited kinetic ion exchange behavior in soils, e.g., aluminum, ammonium, potassium, and several heavy metal cations. According to Ogwada and Sparks (1986), observed kinetic ion exchange behavior was probably due to mass transfer (or diffusion) and chemical kinetic processes. It is postulated that in such 2:1 type minerals, intraparticle diffusion is a rate-controlling mechanism governing the kinetics of adsorption of cations. Jardine and Sparks (1984) showed that the rate of potassium sorption-desorption (in a K-Ca system) was rapid for kaolinite and montmorillonite. However, the rate of potassium exchange was slow for vermiculite. Therefore, we extended our model formulation to account for ion exchange kinetics. The proposed approach was analogous to mass transfer or diffusion between the solid and solution phase such that, for ion species i,

$$\frac{\partial \Gamma_i}{\partial t} = \alpha \left(\Gamma_i^* - \Gamma_i \right) \qquad (1.20)$$

where Γ_i is the amount sorbed on matrix surfaces, Γ_i^* is the equilibrium sorbed amount (at time t), and α is an apparent rate coefficient (d^{-1}) for the kinetic sites. In Equation 1.20, the sorbed amount was calculated using the respective sorption equilibrium condition similar to that given by Equation 1.16. Expressions similar to Equation 1.20 have been used to describe mass transfer between mobile and immobile water as well as chemical kinetics (Parker and Jardine, 1986; Selim et al.,1992). For large α, Γ_i approaches Γ_i^* in a relatively short time and equilibrium is rapidly achieved, whereas for small α, kinetic behavior should be dominant for an extended period of time. Selim et al. (1992) found that incorporating ion exchange kinetics into the competitive transport model described above improved the overall prediction of Cd transport results for two soils compared to the equilibrium ion exchange approach. Improvement in the prediction of the excessive tailing of Cd breakthrough results (or slow release) was realized even though this resulted in lowering the concentration maximum of BTCs.

Specific Sorption

Several sorption and release studies showed that specific sorption mechanisms are responsible for metal ion retention at low concentrations (Garcia-Miragaya and Page,1976; Tiller et al., 1979, 1984a,b). The role of specific sorption and its influence on metal ion behavior in soils has been recognized by several scientists. The general view is that metal ions have a high affinity for sorption sites on oxide-mineral surfaces. Selim et al. (1992) considered the specific sorption process as a kinetic reaction where the rate of sorption is governed by a second-order mechanism. Specific sorption may be assumed to occur between metal ions present in the soil solution and that on specific sites such that

$$\rho \frac{\partial \Psi_i}{\partial t} = k_f \Theta \phi_i C_i - k_b \rho \Psi_i = k_f \Theta \left(S_p - \Psi_i \right) C_i - k_b \rho \Psi_i \tag{1.21}$$

where k_f and k_b are the forward and backward rate coefficients (h[1]), ϕ is the amount of available or vacant specific sites, and Sp is the total amount of specific sorption sites ($mmol_c$ kg^{-1}). Available or vacant specific sites are not strictly vacant. They are assumed to be occupied by hydrogen ions, hydroxyl ions, or by other specifically sorbed species. In the absence of competing metal ions for specific sites (e.g., Ni, Co, Cu, etc.), as is the case in this study, it is reasonable to consider specific sorption as an irreversible process. Therefore, the above second-order reaction was modified to describe irreversible or weakly reversible retention by setting the backward rate coefficient k_b to zero (Selim et al., 1992),

$$\rho \frac{\partial \Psi_i}{\partial t} = k_f \Theta \left(S_p - \Psi_i \right) C_i \tag{1.22}$$

As a result, only two parameters, S_p and k_f, are required to account for irreversible retention. For several metal ions, including Cd, Ni, Co, and Zn, specific sorption has been shown to be dependent on reaction time. Therefore, the use of a kinetic rather than an equilibrium sorption mechanism is recommended. In Equation 1.22, the total amount of specific sites S_p was found to be highly dependent on the type of surface sites and pH (Abd-Elfattah and Wada, 1981; Tiller et al., 1984a,b). Moreover, a major advantage of the formulation of irreversible reaction Equation 1.22 is that a sorption maximum is achieved when all unfilled sites become occupied.

Although model formulation for specific sorption is based on direct reaction between metal ions in soil solution and specific sorption sites, others have considered a consecutive type approach for Cd sorption. Theis et al. (1988) and Selim et al. (1992) considered a set of two second-order reactions: one fully reversible step was followed by an irreversible reaction. Ion exchange as discussed above was not considered. Theis et al. (1988) argued that the amount adsorbed on goethite surfaces was susceptible to migration (via surface diffusion) from primary to secondary surface sites. Other possible mechanisms include hydrolysis of sorbate at the surface and surface precipitation reactions. Earlier, Tiller et al. (1979) quantified specific sorption as the number of sites that retain metal ions following several washings of soil with high concentrations of a nonspecifically sorbed cation (0.01 M Ca_2NO_3). As a result, metal ions on specific sites are not easily replaceable by Ca ions but can be replaced (exchanged) by competing (specifically sorbed) ions such as H^+, Ni^{++}, Cd^{++}, Co^{++}, and Zn^{++}. The relative affinity of specific sorption for such ions was investigated by Abd-Elfattah and Wada (1981) and Tiller et al. (1984a,b). Selim et al. (1992) found that kinetic ion exchange and irreversible specific sorption improved overall predictions of Cd transport in two soils. They also found that the parameter values used to describe specific sorption (k_f) and ion exchange (α) kinetics, which provided improved predictions for single pulse applications, were similar to those used for multiple pulse applications.

COMPETITIVE TRANSPORT

Case Study I — (Cu-Mg)

In this section we present two case studies with emphasis on competitive transport of heavy metals in soils. In the first example, Cu transport was studied using miscible displacement methods where two different soils were studied: Cecil soil was chosen as a bench mark, and McLaren soil. Upon saturation of the soil columns, the fluxes were adjusted to the desired flow rates. A Cu pulse of 100 mg L^{-1} was introduced into each column after it was totally saturated with 0.005 M $MgSO_4$ or $Mg(ClO_4)_2$ as the background solution. Perchlorate as the background solution was used to minimize ion pair formation. The Cu pulse was eluted subsequently with 0.005 M $MgSO_4$ solution. The ionic strength was maintained nearly constant throughout the experiment. In other soil column experiments, similar conditions were used except that no background solution was used in the Cu pulse input solution. This resulted in variable total ionic concentrations or ionic strength during input pulse application and the subsequent leaching solution. A fraction collector was used to collect column effluent.

Figures 1.10 and 1.11 show the effect of total concentration or ionic strength of the input pulse solution on Cu breakthrough results. When Cu was introduced in Mg background solution with minimum change in ionic strength, Cu breakthrough curves (BTCs) appear symmetrical in shape with considerable tailing and a peak concentration of 40 mg L^{-1}. The Mg BTC shows an initial increase in concentration due to slight increase in ionic strength followed by a continued decrease during leaching. When Cu was introduced in the absence of a background solution, the total concentration decreased from 0.005 to 0.0015 M. As shown in Figure 1.11, the Cu BTC showed a sharp increase in concentration due to the chromatographic (or snow-plow) effect (Selim et al., 1992). The peak Cu concentration was 94 mg L^{-1}, and the corresponding Mg concentration in the effluent decreased due to depletion of Mg during the introduction of Cu. The Mg concentration increased thereafter to a steady-state level during subsequent leaching. This snow-plow effect is a strong indication of competitive ion exchange between Mg and Cu cations. The amount of Cu recovered in the effluent was 53% of that applied in the presence of $MgSO_4$ as

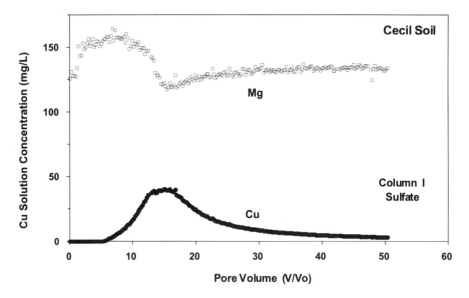

Figure 1.10   Breakthrough results of Cu and Mg for Cecil soil (column I).

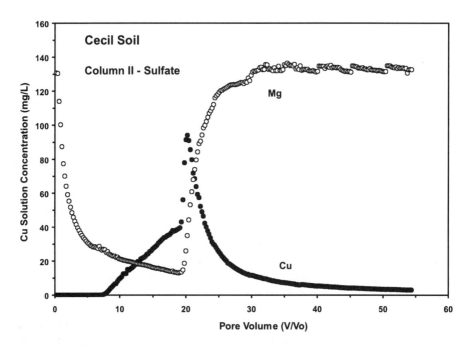

Figure 1.11   Breakthrough results of Cu and Mg for Cecil soil (column II).

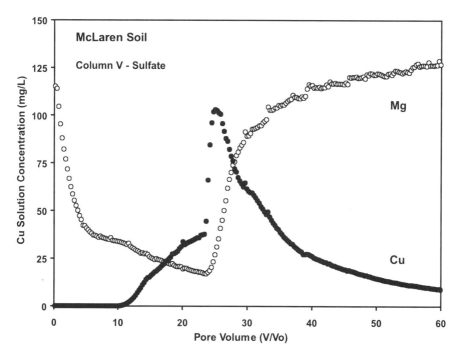

Figure 1.12   Breakthrough results of Cu and Mg for Cecil soil (column V).

the background solution, whereas only 38% was recovered when no background solution was used. For McLaren soil (Figure 1.12), the snow-plow effect was pronounced, as shown in Figures 1.10 and 1.11, due to changes in total concentration of input solutions with a recovery of 60% of that applied. Therefore, miscible displacement experiments indicated that there was strong ion exchange between Cu and Mg cations, which was also affected by the counter ion used. Effluent peak concentrations were three- to fivefold that of the input Cu pulse, which is indicative of pronounced chromatographic effect.

Case Study II — (V-P)

In the second example, we present case studies where the mobility of vanadium (V) and phosphorus (P) is illustrated. Vanadium is one of the metals where anthropogenic input has led to a significant enrichment of the element in the environment. Its mobilization as a consequence of man's industrial activities occurs mainly in the form of particles emitted into the atmosphere. Environmentally, the occurrence of vanadium in petroleum and coal is important because the combustion of these fuels constitutes the major source of vanadium emissions to the atmosphere (Lee, 1983). It has been shown that the burning of fossil fuels mobilizes 12,000 to 24,000 tons of vanadium per year (Bertine and Goldberg, 1971). A large fraction of the vanadium-rich atmospheric particles may enter the soil environment as particulate fallout or dissolved in rain. Vanadate and vanadyl ions are versatile at forming complexes that inhibit or stimulate activity of many enzymes by specific mechanisms.

Phosphorus-soil interactions exhibit various reactions including kinetic, reversible, and nonlinear sorption, often with some degree of irreversibility (Mansell et al., 1992; Chen et al., 1996). The latter primarily occur due to reactions with Fe and Al oxides in the soil. The term *sorption* is used in a general sense here to include P adsorption by mineral surfaces and complexation with humic acids (Al and Fe may form bridges between organic ligands and P ions). Irreversible sorption is considered to involve chemisorption and fixation within mineral structures. Scientists generally agree that many types of P retention can take place simultaneously. Less is known as to whether these reactions operate in series or in parallel. Experimental techniques are available in the literature to monitor sorption and fixation reactions. Although a mathematical model to completely describe P reactions in soil does not exist, models that approximate these reactions do. Multireaction sorption models constitute a class of such approximations and offer a practical means of approximating P mobility during water flow in soils.

In general, the sorption behavior of vanadate in soils is similar to inorganic phosphate. However, V also reduces acid phosphatase activity in soil, which alters the rate of mineralization of organic matter and may reduce phosphate bioavailability (Tyler, 1976). The interactions of V with other ions present in the soil solution (e.g., phosphate) and its potential mobility in the soil profile have not been fully investi- gated (Mikkonen and Tummavuori, 1994). Wang and Selim (1999) carried out a study to quantify the competitive kinetic retention behavior of vanadate and phos- phate in two soils with contrasting properties. Specifically, the focus was to assess the mobility of vanadium in soils where competitive retention is dominant.

Two surface soils with contrasting properties were chosen for this study: a Sharkey clay (very fine, montmorillonitic, nonacid, Vertic Haplaquept) and Cecil soil. The Sharkey soil has an organic matter content of 1.41% and a pH of 5.9, whereas Cecil soil has an organic matter content of 0.74% and a pH of 5.6. Miscible displacement experiments were carried out. Each packed column was slowly satu- rated upward with 0.01 N NH$_4$Cl background solution for 4 d prior to V or P pulse applications. Each pulse contained 100 mg L^{-1} of V or P in 0.01N NH$_4$Cl background solution. An input flow velocity of 0.40 cm h^{-1} using a piston pump was maintained. The effluent from each column was collected using a fraction collector. For each soil, several pulse applications of P, V, and/or P+V were made in a sequential manner. One column received a V pulse followed by a P pulse, then a pulse of a combined P and V solution. Another column received a P pulse followed by a V pulse, then a pulse of a combined P and V solution. In a third column, two or three consecutive pulses of a combined P and V solution were made. Each pulse was approximately 7 pore volumes and was leached using 0.01 N NH$_4$Cl background solution. To test whether equilibrium and/or kinetic V and P retention are the dominant adsorption- desorption processes in soils, a 2 d period of no flow (or flow interruption) during pulse application and/or leaching was carried out.

Miscible displacement results for Cecil soil showed that less P was retained by the Cecil soil than V. This was evident when three consecutive pulses were introduced in the Cecil soil column, as illustrated in Figures 1.13 to 1.15. For the case shown in Figure 1.13, both P and V were present in the input pulse solution during all three

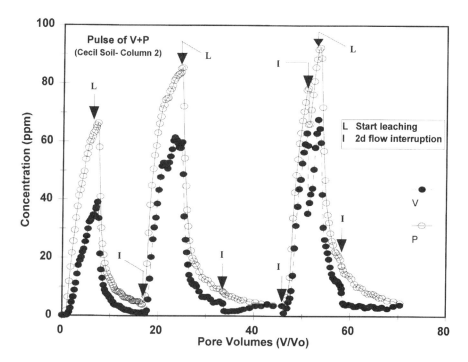

Figure 1.13   Breakthrough results of vanadium (V) and phosphorus (P) from a Cecil soil column that received three consecutive pulse applications. Each pulse contained V and P.

pulses. Based on peak concentration and arrival time for each BTC, Cecil soil had a higher affinity for V than P. This result was consistent for all three consecutive pulses. The location of BTC maximum was the same for both P and V. This suggests that the retention mechanisms for P as well as V on Cecil soil were similar. However, a major difference is that the amount that appears to be irreversibly sorbed (or slowly released) is larger for V than P. We infer the extent of irreversible sorption (or slow release) based on the area under the BTCs. Moreover, the extensive tailing during leaching or desorption also suggests nonlinear and/or kinetic retention mechanisms for both V and P (Mansell et al., 1992).

Competition between V and P for the same retention sites on soil surfaces is evident when one compares BTCs shown in Figure 1.13 with those of Figure 1.14. The absence of P in the pulse solution resulted in a much higher retention of V with a much lower peak concentration when compared to the case when both P and V were present in the input pulse solution. The extent of retention can be estimated from the area under the BTC for V in the first input pulse where four times the amount of V was found in the effluent solution when P was present. In addition, a subsequent P pulse displaced a significant portion of V already retained by the Cecil soil during the first pulse. However, the opposite was not as apparent when a P pulse followed a V pulse, as shown in Figure 1.15.

For all three columns with an input pulse containing both P and V, BTCs for P arrived earlier and with higher peak concentrations than those for V. In addition, a

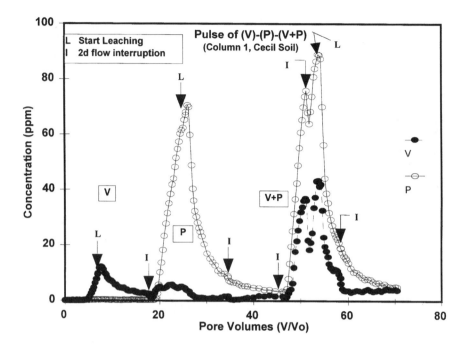

Figure 1.14   Breakthrough results of vanadium (V) and phosphorus (P) from a Cecil soil column that received three consecutive pulse applications. V, P, and a mixture of V plus P were introduced in the first, second, and third pulse, respectively.

decrease in effluent concentration due to flow interruption was observed in all BTCs associated with the third pulse in all three columns, which indicates increases in the amounts of P or V sorbed (Figures 1.13 to 1.15). These observations are consistent for both P and V and show the kinetic nature of the retention mechanisms.

For Sharkey clay soil, miscible displacement results are shown in Figures 1.16 and 1.17 and clearly illustrate the extent of P as well as V retention in this montmorillonitic soil. Unlike the kaolinitic Cecil soil, the extent of P retention by Sharkey soil was stronger than that of V. This is seen in the BTCs associated with the second pulse. As shown in Figure 1.16, the introduction of P pulse was incapable of displacing significant amounts of V from a previous pulse. In contrast, a subsequent P pulse was capable of displacing a significant portion of V as shown in Figure 1.17. We should also emphasize here that based on batch kinetic studies (not shown), the Sharkey soil exhibited very strong retention of V where fast (or near equilibrium) reactions appeared dominant. Although kinetic V sorption continued for 21 d, the amount of rapidly sorbed V (within 24 h) accounted for some 90% of total V sorption. For all input V concentrations, the presence of a constant P concentration (100 mg L^{-1}) resulted in a slight but consistent decrease in the amount of V sorbed at all reaction times. This finding was consistent with that obtained for the kaolinitic Cecil soil, except that the extent of V retention was considerably less than that exhibited by the montmorillonitic Sharkey soil.

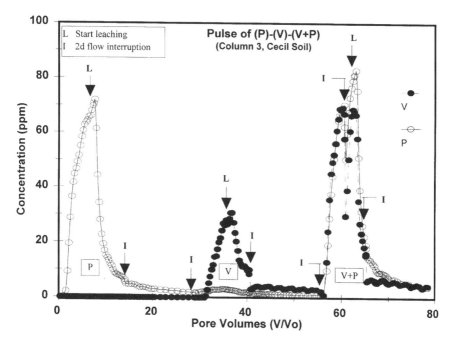

Figure 1.15   Breakthrough results of vanadium (V) and phosphorus (P) from a Cecil soil
column that received three consecutive pulse applications. P, V, and a mixture of
V plus P were introduced in the first, second, and third pulse, respectively.

SUMMARY

We have presented an overview of several models used to describe retention and
release of heavy metals in soils. Major features of multireaction models are that they
are flexible and are not restricted by the number of solute species present in the soil
system nor the governing retention reaction mechanisms. This includes reversible
and irreversible reactions of the linear and nonlinear kinetic types. Moreover, these
models can incorporate concurrent as well as consecutive retention reactions, which
may be at equilibrium or kinetic in nature. Ion exchange mechanisms of the instantaneous and kinetic types were also presented. Case studies of Cu, V, and P isotherms
as well as transport in soil columns provided illustrations of model applications.

  ·   Finally, based on literature review, it was found that most retention experiments
were designed to measure solute sorption, and solute release data were not always
sought. Therefore, kinetic retention models, such as those proposed in this study,
are capable of *predicting* desorption behavior of heavy metals in soils based solely
on adsorption parameters and thus have practical importance. Based on our results,
the overall goodness of fit of our model predictions is considered adequate and lends
added credence to the applicability of the models.

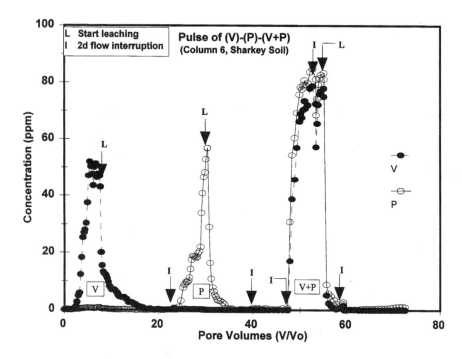

Figure 1.16   Breakthrough results of vanadium (V) and phosphorus (P) from a Sharkey soil
column that received three consecutive pulse applications. V, P, and a mixture of
V plus P were introduced in the first, second, and third pulse, respectively.

REFERENCES

Abd-Elfattah, A. and K. Wada. 1981. Adsorption of lead, copper, zinc, cobalt, and calcium
    by soils that differ in cation-exchange materials. *J. Soil Sci.* 32:271-283.
Ainsworth C. C., J. L. Pilon, P. L. Gassman, and W. G. Van der Sluys. 1984. Cobalt, cadmium,
    and lead sorption to hydrous iron oxide: residence time effect. *Soil Sci. Soc. Am. J.*
    58:1615-1623.
Amacher, M. C., R. W. Brown, R. W. Sidle, and J. Kotuby-Amacher. 1995. Effect of mine
    waste on element speciation in headwater streams. In H.E. Allen et al. (Eds.), *Metal
    Speciation and Contamination of Soil,* Lewis Publishers, Boca Raton. pp 275-309.
Amacher, M. C., J. Kotuby-Amacher, H. M. Selim, and I. K. Iskandar. 1986. Retention and
    release of metals by soils: evaluation of several models. *Geoderma.* 38:131-154.
Amacher, M. C., H. M. Selim, and I. K. Iskandar. 1988. Kinetics of chromium(VI) and
    cadmium retention in soils: a nonlinear multireaction model. *Soil Sci. Soc. Am. J.*
    52:398-408.
Amacher, M. C., H. M. Selim, and I. K. Iskandar. 1990. Kinetics of mercuric chloride retention
    in soils. *J. Environ. Qual.* 19:382-388.
Atanassova, I. D. 1995. Adsorption and desorption of Cu at high equilibrium concentrations
    by soil and clay samples from Bulgaria. *Environ. Pollut.* 87:17-21.
Bittel, J. E. and R. J. Miller. 1974. Lead, cadmium and calcium selectivity coefficients of a
    montmorillonite, illite and kaolinite. *J. Environ. Qual.* 3:250-253.

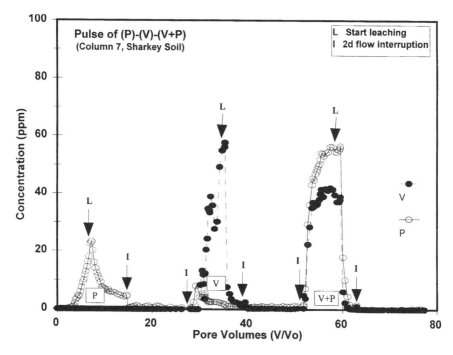

Figure 1.17 Breakthrough results of vanadium (V) and phosphorus (P) from a Sharkey soil column that received three consecutive pulse applications. P, V, and a mixture of V plus P were introduced in the first, second, and third pulse, respectively.

Bertine, K. K. and E. D. Goldberg. 1971. Fossil fuel combustion and the major sedimentary cycle. *Science.* 173:233-235.

Bond, W. J. and R. J. Phillips. 1990. Approximate solution for cation transport during unsteady, unsaturated soil water flow. *Water Resour. Res.* 26:2195-2203.

Buchter, B., B. Davidoff, M. C. Amacher, C. Hinz, I. K. Iskandar, and H. M. Selim. 1989. Correlation of Freundlich K_d and n retention parameters with soils and elements. *Soil Sci.* 148:370-379.

Cavallaro, N. and M. B. McBride. 1978. Copper and cadmium adsorption characteristics of selected acid and calcareous soils. *Soil Sci. Soc. Am. J.* 42:550-556.

Cavallaro, N. and M. B. McBride. 1984. Zinc and copper sorption and fixation by an acid soil clay: effect of selective dissolution. *Soil Sci. Soc. Am. J.* 48:1050-1054.

Chen, J. S., R. S. Mansell, P. Nkedi-Kizza, and B.A. Burgoa. 1996. Phosphorus transport during transient, unsaturated water flow in an acid sandy soil. *Soil Sci. Soc. Am. J.* 60:42-48.

Farley K. J., D. A. Dzombak, and F. M. M. Morel. 1985. A surface precipitation model for the sorption of cations on metal oxides. *J. Colloid and Interface Sci.* 106: 226-242.

Garcia-Miragaya, J. and A. L. Page. 1976. Influence of ionic strength and inorganic complex formation on the sorption of trace amounts of cd by montmorillonite. *Soil Sci. Soc. Am. Proc.* 40:658-663.

Harmsen, K. 1977. Behavior of heavy metals in soils. Centre for Agriculture Publishing and Documentation. Wageningen, The Netherlands.

Harter, R. D. 1984. Curve-fit errors in Langmuir adsorption maxima. *Soil Sci. Soc. Am. J.* 48:749-752.

Hinz, C., B. Buchter, and H. M. Selim. 1992. Heavy metal retention in soils: application of multisite models to zinc sorption. p. 141-170. In I. K. Iskandar and H. M. Selim (Eds.) *Engineering Aspects of Metal-Waste Management.* Lewis Publishers. Boca Raton, FL.

Hinz, C. and H. M. Selim. 1994. Transport of Zn and Cd in Soils: Experimental Evidence and Modelling Approaches. *Soil Sci. Soc. Am. J.* 58:1316-1327.

Hinz, C., H. M. Selim, and L. A. Gaston. 1994. Effect of sorption isotherm type on predictions of solute mobility in soil. *Water Resour. Res.* 30:3013-3021.

Houng, K-H. and D-Y. Lee. 1998. Comparison of linear and nonlinear Langmuir and Freundlich Curve-fit in the study of Cu, Cd, and Pb adsorption on Taiwan soils. *Soil Sci.* 163:115-121.

Jardine, P. M. and D. L. Sparks. 1984. Potassium-calcium exchange in a multireactive soil system. I. Kinetics. *Soil Sci. Soc. Am. J.* 48:39-45.

Jardine, P. M., J. C. Parker, and L. W. Zelazny. 1985. Kinetics and mechanisms of aluminum adsorption on kaolinite using a two-site nonequilibrium transport model. *Soil Sci. Soc. Am. J.* 49:867-873.

Kim Y., R. J. Kirkpatrick, and R. T. Cygan. 1996. [133]Cs NMR study of cesium on the surfaces of kaolinite and illite. *Geochim. Cosmochim. Acta.* 60: 4059-4074.

Kinniburgh, D. G. 1986. General purpose adsorption isotherms. *Environ. Sci. Technol.* 20:895-904.

Lee, K. 1983. Vanadium in the aquatic ecosystem. *Adv. Environ. Sci. Technol.* 13:155-187.

Mansell, R. S., S. A. Bloom, B. A. Burgoa, P. Nkedi-Kizza, and J. S. Chen. 1992. Experimental and simulated P transport in soil using a multireaction model. *Soil Sci.* 153:185-194.

Mikkonen, A. and J. Tummavuori.1994. Retention of vanadium (V) by three Finnish mineral soils. *Eur. J. Soil Sci.* 45:361-368.

Ogwada, R. A. and D. L. Sparks. 1986. Kinetics of ion exchange on clay minerals and soil, I. Evaluation of methods. *Soil Sci. Soc. Am. J.* 50:1158-1162.

Parker, J. C. and P. M. Jardine. 1986. Effect of heterogeneous adsorption behavior on ion transport. *Water Resour. Res.* 22:1334-1340.

Rubin, J. and R. V. James. 1973. Dispersion-affected transport of reacting solution in saturated porous media. Galerkin method applied to equilibrium-controlled exchange in unidirectional steady water flow. *Water Resour. Res.* 9:1332-1356.

Selim, H. M. and M. C. Amacher. 1997. *Reactivity and Transport of Heavy Metals in Soils.* CRC/Lewis, Boca Raton, FL (240 p).

Selim, H. M., B. Buchter, C. Hinz, and L. Ma. 1992. Modeling the transport and retention of cadmium in soils: multireaction and multicomponent approaches. *Soil Sci. Soc. Am. J.* 56:1004-1015.

Sequi, P. and R. Aringhieri. 1977. Destruction of organic matter by hydrogen peroxide in the presence of pyrophosphate and its effect on soil specific surface area. *Soil Sci. Soc. Am. J.* 41:340-342.

Schulte, A. and F. Beese. 1994. Isotherms of cadmium sorption density. *J. Environ. Qual.* 23:712-718.

Sparks, D. L. (ed.). 1999. *Soil Physical Chemistry.* CRC, Boca Raton, FL (409p).

Sposito, G. 1982. On the use of the Langmuir equation in the interpretation of "adsorption" phenomena. II. The "two-surface" Langmuir equation. *Soil Sci. Soc. Am. J.* 46:1157.

Theis, T. L., R. Iyer, and L. W. Kaul. 1988. Kinetic studies of cadmium and ferricyanide adsorption on goethite. *Environ. Sci. Technol.* 22:1032.

Tiller, K. G., J. Gerth, and G. Brümmer. 1984a. The relative affinities of Cd, Ni, and Zn for different soils clay fractions: procedures and partitioning of bound forms and their interpretations. *Geoderma.* 34:1-16.

Tiller, K. G., J. Gerth, and G. Brümmer. 1984b. The relative affinities of Cd, Ni, and Zn for different soils clay fractions and goethite. *Geoderma.* 34:17-35.

Tiller, K. G., V. K. Nayyar, and P. M. Clayton. 1979. Specific and nonspecific sorption of cadmium by soil clays as influenced by zinc and calcium. *Aust. J. Soil Res.* 17:17-28.

Tyler, G. 1976. Influence of vanadium on soil phosphatase activity. *J. Environ. Qual.* 5:216-217.

van Genuchten, M. Th. 1981. Non-equilibrium transport parameters from miscible displacement experiments, Res. Report No. 119, U.S. Salinity Lab., Riverside, Calif. p. 80.

Verburg K. and P. Baveye. 1994. Hysteresis in the binary exchange of cations on 2:1 clay minerals: a critical review. *Clays and Clay Minerals.* 42: 207-220.

Wang, K. and H. M. Selim, 1999. Competitive retention and transport of vanadium in soils. pp. 346-347. *Proc. Fifth Int. Conf. Biogeochem. of Trace Elements,* July 11-15, 1999, Vienna.

Wu, J., D. A. Laird, and M. L. Thompson. 1999. Sorption and desorption of copper on soil clays components. *J. Environ. Qual.* 28:334-338.

CHAPTER 2

Anion and Cation Transport in Zeolitized Tuffs from the Nevada Test Site: Effects of Ion Type, pH, and Ionic Strength

Charalambos Papelis and Wooyong Um

INTRODUCTION

During the period of nuclear weapon production and testing, the U.S. conducted a series of tests at several sites. These tests resulted in contamination from radionuclides as well as from other toxic metals and organic compounds. Common radionuclides found in contaminated areas include uranium and heavy transuranic elements, mainly neptunium, plutonium, and americium, as well as lighter radionuclides, some of which are fission products of the heavier radioisotopes.

In addition to radionuclides, some of the nuclear tests resulted in contamination by other inorganic and organic hazardous substances. These contaminants include the metals lead, copper, and cadmium, anions of arsenic and chromium, and organic contaminants such as polyvinyl chloride (PVC), polystyrene, and phenols (Bryant and Fabryka-Martin, 1991). Some of the above contaminants could be potential health hazards either because of their toxicity or because they are potential carcinogens (Francis, 1994; Sax, 1981).

Assessment of the potential for migration of these contaminants from controlled areas to the accessible environment is based on flow and transport models. These models must account for any possible retardation of the contaminants relative to the groundwater. In addition to hydrologic parameters, therefore, transport models require parameters that reflect the geochemical interactions of contaminants of concern with aquifer materials. The reliability of predictions based on these codes depends directly on the quality of the input parameters. Uncertainties in parameter estimation can lead to significant uncertainties in radionuclide transport simulations, especially given the long time frequently allowed for contaminant migration in model simulations (up to 10,000 years).

Sorbate-sorbent interactions have been commonly incorporated in transport models through the use of an equilibrium distribution coefficient, K_d. Although computationally advantageous, the use of K_d's derived from batch sorption experiments has been criticized for a number of reasons. Use of distribution coefficients assumes that the reactions at the mineral-water interface are always at equilibrium, an assumption referred to as the *local equilibrium assumption* or LEA. Nonequilibrium sorption, however, has been reported in a number of studies, both in the field and in the laboratory (Bahr, 1989; Brusseau et al., 1990; Brusseau et al., 1991; Curtis et al., 1986; Gaber et al., 1995; Goltz and Roberts, 1988; Ma and Selim, 1997; Mackay et al., 1986; Roberts et al., 1986; Selim and Ma, 1995).

Diffusional rate limitations have been frequently used to explain the nonequilibrium sorption of organic chemicals on aquifer materials (Ball and Roberts, 1991; Brusseau et al., 1991; Cunningham et al., 1997; Werth et al., 1997). The slow uptake of inorganic ions by mineral surfaces, however, has been explained by a number of processes, including particle-particle interactions, slow intrinsic reaction rates, and surface precipitation (Davis and Hayes, 1986). The intrinsic chemical reaction rates on nonporous mineral surfaces, however, are believed to be fast (Hayes and Leckie, 1986; Zhang and Sparks, 1990a,b). It is therefore reasonable to assume that the observed slow uptake of inorganic ions by porous or natural materials may also be attributed to diffusional processes, as suggested in a number of studies (Fuller et al., 1993; Papelis et al., 1995; Wood et al., 1990).

A second common objection to the use of distribution coefficients in transport models is based on the methodology used to determine distribution coefficients in the laboratory. Typically, K_d's are determined from sorption experiments with crushed rock or fine particle-size materials; the applicability of parameters derived in this way to the field scale has been questioned. For example, interactions occurring in fractured media, where matrix diffusion may be a significant process, may not be represented adequately by batch sorption experiments. To test the ability of these K_d's to describe organic solute transport under conditions that more closely resemble field conditions, retardation factors obtained from batch sorption experiments have been compared to retardation factors obtained from column experiments (Maraqa et al., 1998, and references therein). Although a number of factors have been suggested to explain the observed discrepancy between retardation factors obtained from batch and from column experiments, none of these factors was able to account for the reported differences (Maraqa et al., 1998).

The scope of this study was to determine retardation factors from sorption of cations and anions in columns of zeolitized tuffs from the Nevada Test Site (NTS) as a function of geochemical parameters. Zeolitized tuffs, produced by the alteration of volcanic materials, are common on the NTS and are considered very good sorbents for radionuclides and other cations (Fawaris and Johanson, 1995; Katz et al., 1996). Because of their high cation exchange capacity, zeolites have been proposed as barriers to prevent the migration of radionuclides and heavy metals (Bailey et al., 1999; Fawaris and Johanson, 1995; Jacobs and Forstner, 1999; Katz et al., 1996). To estimate the ability of zeolitized tuffs to retard the migration of radionuclides at the NTS, column experiments were conducted with anions and cations. Three model cations, lead (Pb^{II}), strontium (Sr^{II}), and cesium (Cs^I), and two anions, chromate

(CrO_4^{2-}, Cr^{VI}) and selenite (SeO_3^{2-}, Se^{IV}), were used to model the behavior of cations and anions with different sorption affinities for different types of sorption sites. Experiments were conducted as a function of pH and ionic strength. The obtained breakthrough curves were evaluated to determine the degree of nonequilibrium. In addition, the obtained retardation factors were compared to retardation factors obtained from batch sorption experiments.

BACKGROUND

Sorption Isotherms

The two major approaches used to model the sorption of radionuclides and other inorganic contaminants on mineral surfaces and soils are surface complexation models and sorption isotherms. Surface complexation modeling has been used successfully to model sorption of both cations and anions on single mineral phases as a function of pH and ionic strength (Hayes and Leckie, 1987; Hayes et al., 1988). Despite the ability of such models to take into consideration the effects of geochemical parameters on ion sorption, ion sorption is still incorporated in most transport codes through the use of sorption isotherms. These are either linear or nonlinear empirical expressions describing the sorption of any sorbate on any sorbent. Although the isotherms are empirical, several of their parameters have a physical, thermodynamic significance. The most commonly used isotherms are briefly described below.

Linear Isotherm

Plotting of sorption data as a linear isotherm results in estimation of a conditional distribution coefficient, K_d ($L^3\,M^{-1}$), a ratio of the mass of sorbate sorbed per mass of sorbent, S ($M\,M^{-1}$), to the aqueous concentration of sorbate in equilibrium with the sorbed contaminant, C_{eq} ($M\,L^{-3}$), as shown in Equation 2.1:

$$K_d = \frac{S}{C_{eq}} \tag{2.1}$$

Distribution coefficients have been used extensively to model organic contaminant sorption on aquifer materials. For inorganic contaminants, however, K_d is frequently a strong function of pH, temperature, and other geochemical conditions (i.e., solution composition, resulting from aqueous reactions and redox potential) (Stumm, 1992). Use of distribution coefficients to model contaminant partitioning at the mineral-water interface assumes that the isotherm is linear and that sorption is controlled by equilibrium (usually referred to as the local equilibrium approach), as opposed to kinetics. Sorption of inorganic contaminants on mineral surfaces is frequently nonlinear. The degree of nonlinearity may be a complex function of the dominant sorption process and other experimental conditions. For example, as the metal ion concentration increases, the onset of surface precipitation may result in increasingly nonlinear isotherms. In addition, distribution coefficients can result in severe errors when used without reference to the specific experimental conditions under which they were determined.

Nonlinear Isotherms

Langmuir Isotherm

The Langmuir isotherm is linear at low surface coverages of the adsorbent but becomes nonlinear at higher surface coverages. The theoretical assumptions behind the Langmuir isotherm are that adsorption occurs only at independent, localized sites with constant sorption energy, independent of surface coverage and that adsorption is limited by the formation of a monolayer (Weber and DiGiano, 1996). The Langmuir isotherm is typically represented by Equation 2.2:

$$S = \frac{Q_a \beta C_{eq}}{1 + \beta C_{eq}} \tag{2.2}$$

where Q_a corresponds to the maximum sorption capacity of the sorbent (M M^{-1}) and β is related primarily to the net enthalpy of adsorption.

Freundlich Isotherm

The Freundlich isotherm is usually represented by Equation 2.3:

$$S = K_F C_{eq}^{1/n} \tag{2.3}$$

where K_F ((M M^{-1})/(M L^{-3})$^{1/n}$) and $1/n$ (–) represent the equivalent of K_d and the exponent of the equilibrium concentration, respectively ($1/n$ is assumed to be 1, by definition, for linear isotherms). The parameter $1/n$ is a function of both the cumulative magnitude and diversity of energies associated with a particular sorption reaction (Weber and DiGiano, 1996). It can also be shown that $1/n$ is related to the enthalpy of adsorption.

Model Description

Several models have been used to simulate the transport of organic and inorganic contaminants (Gaber et al., 1995; Ma and Selim, 1994; Ma and Selim, 1997; Selim and Ma, 1995). The basis of these models is the traditional convection-dispersion equation (CDE). Retardation of solutes, relative to groundwater flow, is incorporated into the models by a retardation factor, R, which reflects the degree of interaction of a solute with aquifer materials. If partitioning can be described by a linear isotherm, the retardation factor, R, is given by Equation 2.4:

$$R = 1 + \frac{\rho K_d}{\theta} \qquad (2.4)$$

where ρ is the bulk density of the medium (ML^{-3}) and θ is the effective porosity of the medium ($L^3 L^{-3}$). If partitioning is described by the nonlinear Freundlich isotherm, the retardation factor is given by Equation 2.5:

$$R = 1 + \frac{\rho}{\theta} \left[K_F \frac{1}{n} C^{(1/n-1)} \right] \qquad (2.5)$$

where C is the aqueous solute concentration (ML^{-3}) and all other parameters are as defined earlier. It is evident by comparing Equations 2.4 and 2.5 that the retardation factor is a nonlinear function of the aqueous solute concentration when solute partitioning is described by a nonlinear isotherm. The simplicity of Equation 2.4, compared to Equation 2.5, explains the commonly used distribution coefficients derived from linear isotherms in transport codes, even if often the observed partitioning behavior is, strictly speaking, not linear.

One of the issues frequently raised concerns the applicability of equilibrium isotherms to model the transport of solutes in soils. The applicability of the local equilibrium assumption (LEA) depends on the relationship between residence time and time required to reach equilibrium. The failure of simple equilibrium models to simulate the transport of contaminants in soil columns and in the field has been attributed to physical or chemical nonequilibrium (Bahr, 1989; Brusseau et al., 1991; Gaber et al., 1995; Goltz and Roberts, 1988; Ma and Selim, 1997; Selim and Ma, 1995).

Transport models have attempted to account for nonequilibrium sorption by incorporating two types of reactive sites and two regions. The two types of sites represent sites that are always in equilibrium with the solute (sites of instantaneous sorption) and sites limited by reaction kinetics. The two regions represent a mobile region, where water movement is controlled by convection and dispersion, and an immobile region. Exchange of solute molecules between the mobile and immobile regions occurs by diffusive processes only.

The limitations of transport modeling based on sorption isotherms should not be ignored. Although these models may be adequate for modeling the transport of organic compounds whose sorption behavior is typically pH independent, the sorption of inorganic ions on mineral surfaces may be a strong function of both pH and ionic strength (Hayes and Leckie, 1987; Hayes et al., 1988). Use of isotherm parameters derived under one set of conditions compared with those derived under different conditions may result in order-of-magnitude errors in the estimation of retardation factors. In addition, sorption of inorganic ions on mineral surfaces may be highly nonlinear. Use of linear isotherms in such cases may also result in substantial errors. For this study, CXTFIT (Parker and van Genuchten, 1984), a CDE model including both equilibrium and nonequilibrium sorption modules was used to model the breakthrough curves (BTC) of all ions. A brief description of the model is given below.

The one-dimensional transport of a single reactive solute under steady-state conditions can be described by Equation 2.6:

$$\frac{\rho}{\theta}\frac{\partial S}{\partial t}+\frac{\partial C}{\partial t}=D\frac{\partial^2 C}{\partial x^2}-v\frac{\partial C}{\partial x}-\mu_w C-\frac{\mu_s \rho}{\theta}S+\gamma_w+\frac{\gamma_s \rho}{\theta} \qquad (2.6)$$

where C is the volume-averaged resident concentration of the solute in the liquid ($M\,L^{-3}$), S is the adsorbed concentration per unit mass of the solid phase ($M\,M^{-1}$), x is the distance (L), t is the time (t), D is the dispersion coefficient ($L^2\,t^{-1}$), v is the average pore-water velocity ($L\,t^{-1}$), ρ is the porous medium bulk density ($M\,L^{-3}$), θ is the effective porosity of the medium ($L^3\,L^{-3}$), μ_w and μ_s are the rate constants for first-order decay in the liquid and solid phase, respectively (t^{-1}), and γ_w and γ_s are the rate constants for zero-order products in the liquid and solid phase ($M\,L^{-3}\,t^{-1}$).

If linear equilibrium is assumed to describe solute sorption adequately, the above equation can be simplified to the following Equation 2.7:

$$R\frac{\partial C}{\partial t}=D\frac{\partial^2 C}{\partial x^2}-v\frac{\partial C}{\partial x}-\mu C+\gamma \qquad (2.7)$$

where R is the retardation factor (–),

$$\mu=\mu_w+\mu_s\rho\frac{K_d}{\theta} \qquad (2.8)$$

the new rate constant, γ, is given by Equation 2.9:

$$\gamma=\gamma_w+\gamma_s\frac{\rho}{\theta} \qquad (2.9)$$

and all other parameters are as defined above.

The two-site two-region model (TSM/TRM) can be thought of as a combination of a two-site model (TSM) and a two-region model (TRM). The TSM assumes that sorption on type-1 sites is always at equilibrium (i.e., instantaneous sorption), whereas sorption on type-2 sites is rate limited. The governing equations for the two types of sites are given by Equations 2.10 and 2.11:

$$\frac{\partial S_1}{\partial t}=fk_1\frac{\partial C}{\partial x} \qquad (2.10)$$

$$\frac{\partial S_2}{\partial t}=a\left(k_2 C-S_2\right) \qquad (2.11)$$

where f is the fraction of type-1 sites, k_1 is an equilibrium constant ($L^4\,M^{-1}\,t^{-1}$), α is a kinetic rate constant (t^{-1}), and k_2 is another equilibrium constant ($L^3\,M^{-1}$). If production and decay terms are ignored, the transport equation can be represented by

$$\beta R \frac{\partial C_1}{\partial T} + (1-\beta)R\frac{\partial C_2}{\partial T} = \frac{1}{P}\frac{\partial^2 C_1}{\partial Z^2} - \frac{\partial C_1}{\partial Z} \tag{2.12}$$

where

$$(1-\beta)R\frac{\partial C_2}{\partial T} = \omega(C_1 - C_2) \tag{2.13}$$

and

$$\beta = \frac{\theta_m + F\rho K_d}{\theta + \rho K_d} \tag{2.14}$$

$$\omega = \alpha(1-\beta)RL/v \tag{2.15}$$

where α is a first-order rate constant (t^{-1}), θ_m is the porosity associated with type-2 sites, F represents the fraction of sites occupied by type-1 sites, P is the Peclet number, a dimensionless ratio of convection and dispersion forces given by $P = vL/D$, T is dimensionless time ($T = vt/L$), and Z is dimensionless distance ($Z = x/L$).

Finally, according to the TRM, two types of pore spaces exist in a soil column. In the first type, the processes controlling solute transport are convection and dispersion, while in the second type of pore spaces, the controlling process is diffusion. The first region is referred to as the mobile region, while the second is referred to as the immobile region. The governing equations for this model are given by Equations 2.16 and 2.17:

$$\left(\theta_m + f\rho K_d\right)\frac{\partial C_m}{\partial t} + \left[\theta_{im} + (1-f)\rho K_d\right]\frac{\partial C_{im}}{\partial t} = \theta_m D_m \frac{\partial^2 C_m}{\partial x^2} - q\frac{\partial C_m}{\partial x} \tag{2.16}$$

$$\left[\theta_m + (1-f)\rho K_d\right]\frac{\partial C_{im}}{\partial t} = \alpha*\left(C_m - C_{im}\right) \tag{2.17}$$

where q is the Darcy velocity (L t^{-1}) and α^* is a first-order rate constant. Because the dimensionless transport equation and the initial and boundary conditions are identical for the TSM and TRM, parameters obtained by fitting the BTC data to the TSM can also be interpreted in terms of the TRM. Alternatively, one could consider the observed nonequilibrium as resulting either from a rate-limited sorption reaction, or a diffusional process in the immobile region.

MATERIALS AND METHODS

Material Characterization

Zeolitized tuffs are present in the lower half of the tuffaceous section exposed at Rainier Mesa of the Nevada Test Site. They compose the lower half of the Grouse Canyon Member of the Indian Trail Formation and bedded tuff of the Paintbrush

Tuff. These units are typically 270 to 400 m in thickness. The samples used in this study were collected from Tunnel Bed 3, a section of the local informal units of the Indian Trail Formation. The complete characterization of the material was reported elsewhere (Sloop, 1998). Only a summary of the characterization is given here. The zeolitized tuff was broken down into smaller pieces for the column experiments. The material was sieved and experiments were conducted with two size fractions: particles with diameters between 0.85 and 1.18 mm and particles with diameters between 2.85 and 4.00 mm. All but one of the column experiments were performed with the larger size fraction. The true density of the material was 2.32 g cm^{-3}, consistent with expected densities for zeolites and feldspars (Klein and Hurlbut, 1993). The intraparticle porosity was 0.20. Bulk densities and interstitial porosities were determined specifically for each column and are reported in the Results and Discussion section.

The specific surface area of the adsorbent was measured with a Micromeritics Gemini 2370 surface area analyzer using nitrogen adsorption and the BET model (Brunauer et al., 1938). The specific surface areas measured were 10.64 and 8.13 m^2 g^{-1} for the smaller and larger size fractions, respectively. In addition, the surface area of much finer particles with a mass median diameter of 20 μm was 13.63 m^2 g^{-1}. These results indicate very little dependence of the surface area on size, suggesting that the majority of the surface area is internal and accessible. The external surface area of smooth spherical particles with 4.00 mm diameter and 2.32 g cm^{-3} particle density is 6.5×10^{-4} m^2 g^{-1}, orders of magnitude lower than the measured surface area, even if surface roughness is taken into account. Values on the order of 10 m^2 g^{-1} are therefore a clear indication of substantial and accessible porosity. These results have significant implications for the sorption capacity of the adsorbent and will be discussed further in the Results and Discussion section.

The mineralogy of the adsorbent was determined using a variety of techniques. Bulk mineralogy was determined by bulk X-ray diffraction (XRD). Based on XRD, the major components of the sample were clinoptilolite (a zeolite) and feldspars (albite, anorthite, plagioclase). Based on the elemental composition of the tuff, as determined by X-ray fluorescence (XRF), and the reported groundwater composition (a sodium bicarbonate water), it can be inferred that the clinoptilolite was mostly Na-saturated.

Additional, semiquantitative mineralogical composition of the core was obtained from scanning electron microscopy (SEM) combined with energy dispersive X-ray (EDX) spectroscopy. This analysis was performed on microprobe-polished thin sections of the rock and combined with image analysis. The image analysis is based on the intensity of elastically backscattered electrons as a function of the atomic number of the backscattering element. Heavier elements are more efficient backscatterers, so that minerals containing heavier elements (e.g., iron oxides) appear brighter than minerals containing lighter elements (e.g., quartz). By identifying the composition of grains of a particular gray scale (using EDX) and determining the percentage of the thin section corresponding to the corresponding gray scale, one can obtain at least a semiquantitative estimate of the mineralogical composition of a rock, including disordered and amorphous phases which may not be detectable by

XRD. One limitation of the technique is that because of similar composition, different minerals may sometimes not be distinguishable based on backscatter image intensity and EDX. Albite and hydrated aluminosilicates (zeolites), for example, are sometimes indistinguishable, based on their composition and therefore backscattering intensity alone.

We were able to distinguish between hydrous aluminosilicates (zeolitic material) and feldspars based on the damage caused by the electron beam. Evaporation of the water in the zeolitic material caused by the energy of the focused beam is clearly visible after a few minutes of exposure; feldspars, on the other hand, are essentially unaffected by the electron beam. The SEM/EDX analysis results were consistent with the XRD analysis. Based on the above combination of analyses, the sample appeared to contain approximately equal quantities (45 to 50% by weight) of clinoptilolite and feldspars (albite, anorthite, and plagioclase). In addition to the major phases (clinoptilolite and feldspars), mordenite (a zeolite), and quartz were also present in the sample. Finally, biotite and iron and titanium oxides were also present in small amounts (approximately 2.5% by weight).

Column Experiments

All column experiments were conducted in glass chromatographic columns (Fisher Scientific, NJ). The length of the columns was 15 cm, and the inner diameter 1.5 cm. Filters with 0.45 µm nominal pore diameter were placed at both ends of the column to minimize adsorbent loss during the experiment. Both particle size fractions were used during these experiments (0.85–1.18 and 2.85–4.00 mm). The wet packing method was used for packing all columns (Oliviera et al., 1996). This method is recommended for packing columns for saturated flow conditions, because it minimizes the possibility of air being trapped in the interstitial pore space between particles.

The solid particles were saturated with either NANOpure™ water or a $NaNO_3$ (sodium nitrate) solution of the desired ionic strength before packing the column. The fully saturated particles were then introduced into the column, which was already filled with the background electrolyte used in the experiments ($NaNO_3$), through a funnel, while vibrating the column. The pH of the solution was adjusted before packing the column, by adjusting the pH of the solution used to saturate the solid particles by addition of either acid or base until the intended solution pH was reached. The effluent pH was monitored during the experiments; because of the high buffer capacity of the system (due to the high solid concentration), the pH of the effluent was never significantly different from the pH of the influent.

The adsorbate solution was introduced from the bottom of the column to ensure saturated flow conditions. For all runs, a step-function increase in adsorbate concentration was used. A concentration of either 10^{-5} or 10^{-4} M was employed in all experiments. The velocity was regulated by a constant head reservoir. By adjusting the height of the reservoir, the flow rate in the columns could be adjusted. Based on the experimentally measured flow rate and the hydraulic head, the hydraulic conductivity in the column could be experimentally determined.

The bulk density of the adsorbent was determined separately for each column from the total dry weight of the solids and the total volume of the column. The total porosity of the column, including both intraparticle porosity and interstitial porosity, was calculated by the following equation:

$$\varepsilon = 1 - \frac{\rho_b}{\rho_t} \tag{2.18}$$

where ρ_b is the bulk density, estimated as described above, and ρ_t is the true density of the sorbent, as reported in the Adsorbent Characterization section. It should be emphasized that the porosity determined as described above is the total porosity of the column, including both the interparticle (interstitial or effective) porosity and intraparticle (internal) porosity. Only the interstitial (effective) porosity is important with respect to solute flow, whereas the intraparticle porosity is controlling intraparticle diffusion. The intraparticle porosity was estimated from mercury porosimetry measurements (Gregg and Sing, 1982). At approximately atmospheric pressure, mercury cannot penetrate any pores but the largest, so the density determined by mercury porosimetry can be used as an approximation of the envelope density (mass of solid particles divided by the volume of the particles, including intraparticle pore spaces). The intraparticle porosity can then be estimated by

$$\varepsilon_{int} = 1 - \frac{\rho_{env}}{\rho_t} \tag{2.19}$$

where ρ_{env} is the envelope density determined by mercury porosimetry. Finally, the average linear velocity through the column, v_x ($L\ t^{-1}$), can be estimated from the volumetric flow rate and the effective porosity. Similarly, the total volume of solution discharged can be expressed in terms of pore volumes, where one pore volume is the volume of water corresponding to the effective porosity of the column.

Analytical Methods

The cations used in the experiments were lead, strontium, or cesium, and the anions were chromate or selenite. The adsorbates were added as either $Pb(NO_3)_2$ (lead nitrate), $CsNO_3$ (cesium nitrate), $Sr(NO_3)_2$ (strontium nitrate), K_2CrO_4 (potassium chromate), or Na_2SeO_3 (sodium selenite) depending on the experiment performed. All reagents used were of ACS reagent grade quality or better. High resistivity NANOpure water was used exclusively for all solutions. The samples were analyzed using a Perkin Elmer 4110 ZL atomic absorption spectrophotometer with a graphite furnace and Zeeman background correction. Duplicates were run for each sample and the results were averaged. The relative standard deviation of the duplicate measurements was typically less than 5%. In all cases, a peak area mode was used for analysis using appropriate metal standards.

Table 2.1 Experimental Conditions for Column Experiments

Sample I.D.	Particle Nominal Size (mm)	Metal Concentration (M)	Ionic Strength (M)	pH
Cr-9	0.85 – 1.18	10^{-5}	0.01	7.68
Cr-6	2.85 – 4.0	10^{-5}	0.01	6.76
Se-1	2.85 – 4.0	10^{-5}	0.01	7.42
Pb-3	2.85 – 4.0	10^{-5}	0.01	7.45
Pb-4	2.85 – 4.0	10^{-5}	1.0	3.15
Sr-2	2.85 – 4.0	10^{-5}	0.01	7.47
Sr-6	2.85 – 4.0	10^{-4}	1.0	8.90
Cs-1	2.85 – 4.0	10^{-5}	1.0	8.10

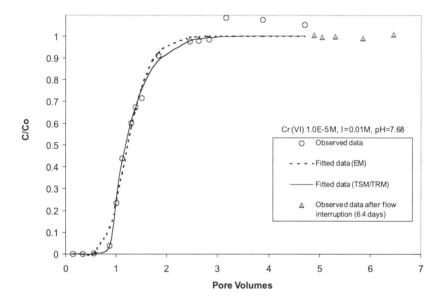

Figure 2.1 Breakthrough curve of Cr^{VI}; I = 0.01 M; pH = 7.68.

RESULTS AND DISCUSSION

Anion Column Experiments

Chromate Sorption

Column experiments with anions will be discussed first, followed by a discussion of experiments with cations. The list of experiments discussed here is not exhaustive; however, it is representative in terms of metal concentration, ionic strength, and pH. The experimental conditions are shown in Table 2.1 and the modeling parameters and results are shown in Table 2.2. The breakthrough curve of 10^{-5} M Cr^{VI} in 0.01 M ionic strength at pH 7.68 is shown in Figure 2.1.

Table 2.2   Summary of Modeling Parameters and Results

Sample I.D.	Bulk Density (ρ_b) (g cm^{-3})	Effective Porosity (cm^3 cm^{-3})	Velocity (v_x) (cm d^{-1})	R (EM)	D (EM) (cm^2 d^{-1})	R (TSM/TRM)	D (TSM/TRM) (cm^2 d^{-1})	β	ω
Cr-9	0.86	0.51	72.8	1.28	4.4	1.32	3.3	0.74	1.39
Cr-6	0.63	0.64	1295.2	1.13	1520.8	1.18	145.8	0.69	1.04
Se-1	0.64	0.63	4765.9	1.13	9718.4	2.29	2181.7	0.40	0.20
Pb-3	0.7	0.60	729.3						
Pb-4	0.7	0.61	239.8	165.6	4649.2	116.3	116.7	0.01	2.50
Sr-2	0.67	0.62	460.5						
Sr-6	0.71	0.59	485.3	1.90	4630.8	1.97	778.8	0.51	0.42
Cs-1	0.7	0.61	820.1	106.2	5189.7	97.9	8002.0	0.02	3.85

R: retardation factor; D: dispersion coefficient; EM: equilibrium model; TSM: two-site model; TRM: two-region model.

The effluent Cr^{VI} concentration is shown as a relative concentration (C/Co) and as a function of the total volume eluted relative to one column pore volume. For conservative tracers, the midpoint of the relative concentration (C/Co = 0.5) is expected to appear at exactly one pore volume. The overall solute retardation can therefore be estimated, at least semiquantitatively, by inspection of the breakthrough curve (Freeze and Cherry, 1979). The retardation factors obtained based on both the equilibrium model (EM) and the two-site/two-region model (TSM/TRM) are shown in Table 2.2. Both models could describe the observed behavior fairly well (Figure 2.1).

The retardation factors obtained, 1.28 and 1.32 for the EM and TSM/TRM, respectively, suggest very little retardation for this anion under these conditions. These results are consistent with the low anion exchange capacities of zeolites reported in the literature (Ming and Mumpton, 1989) and with previous batch equilibrium sorption experiments performed in our laboratory (Sloop, 1998). The only studies reporting any appreciable sorption of anions by zeolites were conducted with surfactant modified zeolites (Haggerty and Bowman, 1994; Jacobs and Forstner, 1999; Li and Bowman, 1997). The batch laboratory results indicated no sorption under neutral or above neutral pH, even at relatively high solid concentrations (100 g L^{-1}). The expected retardation factor was therefore 1. The even higher solid concentration in the column experiments and diffusion of the solute in intraparticle pore spaces could account for the small retardation factor increase in the column experiments compared to the batch sorption experiments. In the field, the equivalent phenomenon of retardation because of matrix diffusion is well documented (Freeze and Cherry, 1979). It should be remembered that the intraparticle porosity of the zeolitized tuff particles was 0.20. Retardation because of diffusion of solute in intraparticle pore spaces is therefore possible. An additional uncertainty in estimating retardation factors is caused by the uncertainty in the interstitial pore volume estimate. The interstitial pore volume estimate was obtained from the porosity of the zeolitized tuffs, as determined by mercury porosimetry, and is therefore operationally defined.

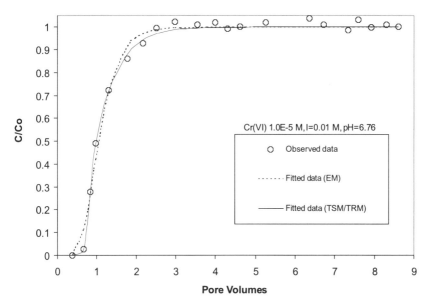

Figure 2.2 Breakthrough curve of Cr[VI]; I = 0.01 M; pH = 6.76 (larger size fraction).

The breakthrough curve appeared to be symmetric. Tailing of breakthrough curves has been interpreted as an indication of nonequilibrium sorption. To test the hypothesis of equilibrium vs. nonequilibrium sorption, the flow interruption technique was used (Brusseau et al., 1989; Reedy et al., 1996). The technique consists of interrupting the steady flow of solute for a specified time to allow equilibrium to be reached. The effluent concentrations before and after the interruption are compared. Lower effluent concentration after resuming flow, compared to the concentration before the flow interruption, is an indication of nonequilibrium sorption, either chemical or physical. Inspection of Figure 2.1 suggests that nonequilibrium sorption was most likely not important under these conditions, even at the high linear velocity used in this experiment (72.8 cm d^{-1}). This is consistent with the minimum Cr[VI] sorption under these conditions. In summary, these results suggest only marginal retardation of Cr[VI] through zeolitized tuffs at a pH value similar to the pH of the groundwater at the NTS.

Experiments were also conducted with larger particles, having nominal diameters between 2.85 and 4.0 mm, and the results are shown in Figure 2.2. All other conditions were approximately the same, the pH being slightly lower, 6.76. The breakthrough curve was again fairly symmetrical, despite the very high average linear velocity (1295 cm d^{-1}). The higher linear velocity was a result of the larger particle size used in this experiment. Although flow interruption data are not shown, sorption nonequilibrium was not considered likely, given the very low affinity of Cr[VI] for these surfaces.

It can be seen from Table 2.2 that the estimated retardation factors were again negligible for practical purposes and probably indistinguishable from 1, and essentially

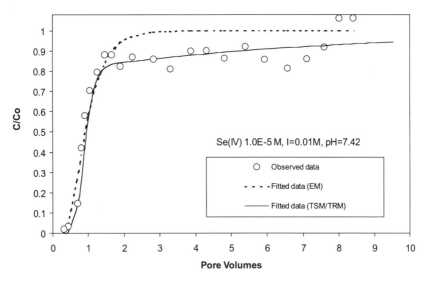

Figure 2.3   Breakthrough curve of Se^{IV}; I = 0.01 *M*; pH = 7.42.

the same for the EM and TSM/TRM, 1.13 and 1.18, respectively. Both the EM and the TSM/TRM retardation factors were smaller compared to the retardation factors obtained with the smaller size fraction. The slight decrease in the retardation factor could be a function of the larger size fraction. It should be expected that as the particle size increases, the specific surface area should decrease and, therefore, the K_d's and retardation factors should also decrease. In addition, pore spaces in the column with larger size particles should be more easily accessible, leading to lower retardation factors because of diffusion. The relatively weak dependence of specific surface area on particle size is consistent with the relatively weak dependence of estimated retardation factors on particle size.

Selenite Sorption

The breakthrough curve of 10^{-5} *M* Se^{IV} in 0.01 *M* ionic strength at pH 7.42 is shown in Figure 2.3. Very little retardation was observed for this anion as well. A slow increase in concentration past the initial breakthrough, however, could be observed. This behavior could be attributed to sorption nonequilibrium, given the very high flow rate used in this experiment. Specifically, the flow rate used in this column experiment, 4765.9 cm d^{-1}, was more than two orders of magnitude higher than estimated flow rates in vertical fractures in tuffs of Yucca Flat on the NTS (approximately 20 cm d^{-1}). Even under such fast flow conditions, however, we believe that sorption was not limited by the intrinsic chemical reaction rate. Sorption reactions have been shown to be fast on nonporous oxide surfaces (Hayes and Leckie, 1986; Zhang and Sparks, 1990a,b). The observed nonequilibrium behavior is there-fore believed to be caused by the much slower, compared to chemical reaction, diffusion of selenite in intraparticle pore spaces. The retardation factors estimated from the EM and TSM/TRM were 1.13 and 2.29, respectively (Table 2.2).

These estimates are in fairly good agreement with the batch sorption experiments. Although batch experiments at exactly these conditions were not available (I = 0.01 M, pH = 7.4), the retardation factor estimated based on the K_d determined at pH 7 and 1.0 M ionic strength was 2.14. Given the proximity of pH values and the relatively minor ionic strength dependence of Se^{IV} sorption (Hayes et al., 1988; Sloop, 1998), the estimated retardation factor from batch sorption experiments could probably be used to describe sorption in the column experiments as well. A comparison of the Cr^{VI} and Se^{IV} breakthrough curves suggests a slightly higher retardation of Se^{IV} compared to Cr^{VI}. This result is consistent with previously reported sorption data with the two anions on several mineral surfaces and previous experimental results with zeolitized tuffs from the NTS (Sloop, 1998). In conclusion, neither of the anions examined as part of this study is expected to be retarded significantly during transport through zeolitized tuffs.

Cation Column Experiments

Lead Sorption

The high cation sorption capacity of zeolites is well known and has been reported (Bailey et al., 1999; Barrer, 1978; Breck, 1974; Dyer, 1995; Dyer et al., 1991; Jacobs and Forstner, 1999; Loizidou and Townsend, 1987; Malliou et al., 1994). In addition, Pb is one of the most strongly binding cations reported in selectivity series with respect to several minerals, including amphoteric oxides and hydroxides, as well as permanent charge clays and zeolites (Stumm, 1992). Very high sorption of Pb on zeolitized tuffs from the NTS was previously measured in our laboratory (Sloop, 1998). Specifically, under certain conditions, sorption of Pb was so high that no K_d could be estimated with any reasonable reliability. The column experiments confirmed this sorption behavior.

Relative effluent concentrations as a function of pore volumes eluted are shown in Figure 2.4, representing sorption of 10^{-5} M Pb in 0.01 M ionic strength, at pH 7.45. No significant Pb breakthrough occurred even after more than 1800 pore volumes. As mentioned earlier, the exact retardation factor cannot be estimated from batch sorption experiments under these experimental conditions, but it was on the order of thousands. The results of the column experiments, therefore, are consistent with the batch sorption experiments.

The experimental conditions used during this run were similar to geochemical conditions expected at the NTS. The pH was slightly below the expected pH of the groundwater at the NTS (approximately 8) and the ionic strength (0.01 M) was also higher than measured in NTS groundwater (approximately 0.003). The observed retardation should therefore represent a conservative estimate of Pb transport through zeolitized tuffs at the NTS. The nature of Pb sorption under these conditions is not clear. Based on sorption experiments performed in our laboratory as a function of pH and ionic strength, it appears that Pb sorption under these conditions may be a combination of ion exchange and surface precipitation (Bernot, 1999; Sloop, 1998).

The formation of Pb polynuclear complexes and surface precipitates on oxides has been documented (Bargar et al., 1998; Chisholm-Brause et al., 1990; Chisholm-Brause

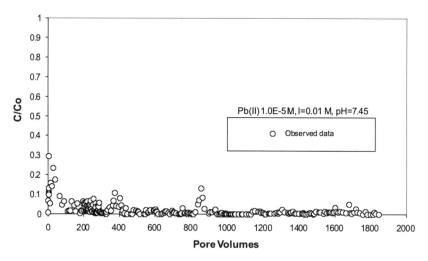

Figure 2.4   Effluent concentration of Pb as a function of pore volumes eluted; I = 0.01 M; pH = 7.45.

et al., 1989; Roe et al., 1991). Increased pH would increase the potential for Pb hydrolysis and therefore the potential for surface precipitation (Baes and Mesmer, 1986). Similarly, Pb would be increasingly excluded from internal ion-exchange sites of zeolites with increasing ionic strength (because of increased competition with the background electrolyte cation for permanent charge sites) thereby increasing the potential for formation of surface polynuclear complexes and precipitates. These effects have been reported for Co sorption on ion exchange sites of smectite clays (Papelis and Hayes, 1996). The tendency of Pb to form surface precipitates would also increase with increasing total Pb concentration. It is therefore possible that at concentrations substantially lower than 10^{-5} M, surface precipitates would not form. In that case, however, the high cation exchange capacity of these tuffs would still result in very high Pb retardation.

To isolate the influence of cation exchange processes on Pb retardation, experiments were also performed at higher ionic strength (1.0 M) and low pH (3.15). The higher Na concentration would tend to reduce Pb sorption on cation exchange sites of the zeolites, and the lower pH would minimize Pb hydrolysis and therefore the potential for the formation of surface precipitates. It should be emphasized that at 1.0 M ionic strength, the Na cation concentration would be five orders of magnitude higher than the concentration of Pb (assuming Pb concentration 10^{-5} M) so that Pb sorption would be substantially reduced, despite the higher affinity of Pb for cation exchange sites compared to Na. In addition, sorption at low pH would minimize the contribution of amphoteric oxides on the sorption capacity of the tuff. The corresponding breakthrough curve is shown in Figure 2.5. Modeling of the data with the EM and TSM/TRM resulted in retardation factors of 165.6 and 116.3, respectively. Obviously, the TSM/TRM can simulate the experimental data better, especially during the initial phase of the experiment.

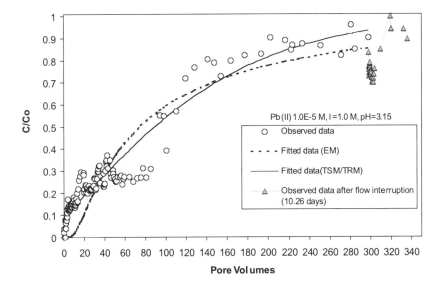

Figure 2.5 Breakthrough curve of Pb; I = 1.0 *M*; pH = 3.15.

The substantially better fit of the data to the TSM/TRM suggests that sorption nonequilibrium may be present. This would not be surprising, given the high linear velocity during the experiment (240 cm d^{-1}). To test this hypothesis, the flow was interrupted after approximately 300 pore volumes and resumed approximately 10 days later. The initial decrease in concentration after flow resumption is a clear indication of nonequilibrium sorption. The retardation factor obtained from TSM/TRM agreed fairly well with batch sorption experiments (Bernot, 1999; Sloop, 1998). These values were 116.3 and 95.2 for the TSM/TRM and batch equilibrium experiments, respectively.

Strontium Sorption

Normalized Sr effluent concentrations as a function of pore volumes eluted are shown in Figure 2.6. The ionic strength was 0.01 *M* and the pH was 7.47. It can be seen from Figure 2.6 that no significant breakthrough occurred even after elution of more than 2000 pore volumes. These results suggest very high retardation of Sr under these experimental conditions. It should be emphasized that the conditions represented in this column experiment could be characterized as conservative with respect to the geochemical conditions at the NTS. This is because the ionic strength of NTS groundwater is less than 0.01 *M*, and Sr is expected to be retarded even more under lower ionic strength conditions. These data are also in good agreement with the retardation factor derived from batch equilibrium sorption experiments, which was 2519 (Sloop, 1998).

The breakthrough curve for 10^{-4} *M* Sr in 1.0 *M* ionic strength background electrolyte at pH 8.9 is shown in Figure 2.7. Rapid breakthrough was observed under

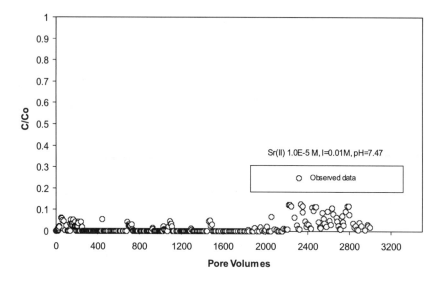

Figure 2.6   Effluent concentration of Sr as a function of pore volumes eluted; I = 0.01 *M*; pH = 7.47.

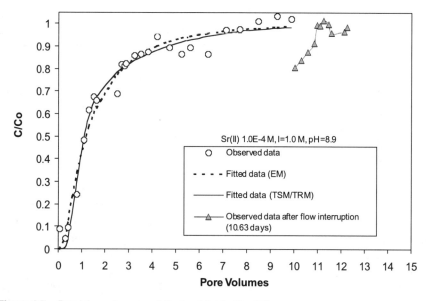

Figure 2.7   Breakthrough curve of Sr; I = 1.0 *M*; pH = 8.9.

these conditions and retardation factors obtained by modeling the data with the EM and TSM/TRM were 1.90 and 1.97, respectively. The much faster breakthrough compared to the column experiment shown in Figure 2.6 is primarily due to the 100-fold increase in ionic strength. Under the high ionic strength conditions, Sr cannot compete with background electrolyte cations for ion exchange sites, and

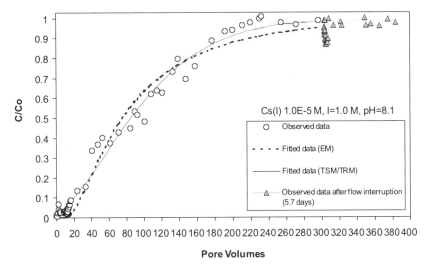

Figure 2.8   Breakthrough curve of Cs; I = 1.0 *M*; pH = 8.1.

sorption is severely limited. These low retardation factors agree reasonably well with the batch sorption data. Under similar conditions, no Sr sorption was observed at the low solid concentrations used in the batch sorption experiments, corresponding to a retardation factor of 1 (Sloop, 1998). It is possible that some limited sorption would be observed at the much higher solid concentrations used in the column experiments. The estimated retardation factors of less than 2 agree fairly well with this hypothesis and an increased apparent retardation because of diffusion in pore spaces.

A flow interruption at approximately 10 pore volumes for approximately 10 days revealed nonequilibrium sorption for these conditions as well. This is not surprising given the high average linear velocity during this experiment (485 cm d^{-1}) and the large size of the adsorbent particles used. Batch rate of uptake experiments with Sr suggest equilibrium time on the order of 20 days for this particle size (Bernot, 1999). It should be kept in mind that Sr is unlikely to hydrolyze to any significant extent, so that the observed sorption and retardation should be attributed to only cation exchange reactions.

Cesium Sorption

The breakthrough curve of Cs as a function of pore volumes eluted is shown in Figure 2.8. The data correspond to sorption of Cs in 1.0 *M* ionic strength background electrolyte at pH 8.1. The retardation factors obtained from the EM and TSM/TRM were 106.2 and 97.9, respectively. These results appear to be in good agreement. These retardation factors are lower than predicted from batch equilibrium sorption experiments (*R* = 188) (Sloop, 1998). Differences in retardation factors estimated from batch and column experiments have been reported in the literature (Maraqa et al., 1998, and references therein). Typically, batch experiments tend to overestimate retardation factors compared to column experiments. Reasons given for this

frequently observed discrepancy include sorption nonsingularity, sorption nonequilibrium, presence of immobile water regions in the column, reduction of particle
spacing in the column, and sorbent loss through the column (Maraqa et al., 1998).

Sorption nonequilibrium could be suspected by inspection of the Cs breakthrough
curve and the fact that the retardation factor from the batch experiments was approximately 190. Decreasing C/Co after flow interruption for approximately 6 days is
consistent with nonequilibrium sorption under these experimental conditions. A
similar situation, however, existed during other experiments and did not result in
differences in retardation factors of the same magnitude (almost a factor of 2). It
should also be remembered that the batch equilibrium sorption experiments were
conducted with the smaller size fraction. Although the specific surface area difference
between the two size fractions was not dramatic, the K_d's, and therefore retardation
factors, in the batch and column experiments should reflect these differences. Given
all uncertainties mentioned, the agreement in retardation factors from batch equilibrium and sorption experiments could be considered fair.

SUMMARY AND CONCLUSIONS

Column experiments were conducted to study the sorption of cations (Pb, Sr,
and Cs) and anions (Cr^{VI} and Se^{IV}) on zeolitized tuffs from the NTS. Column
experiments have traditionally been used, as an alternative to batch sorption experiments, to obtain a more realistic estimate of solute transport in aquifers. Based on
previous batch sorption experiments, it was determined that sorption may be a
function of pH and ionic strength, depending on the solute. Experiments were
therefore conducted under different pH and ionic strength conditions. Most experiments were performed at neutral or slightly basic pH, to represent geochemical
conditions of the groundwater at the NTS. Similarly, most experiments were conducted at 0.01 M ionic strength, which is not much higher than the ionic strength
of the groundwater at the NTS. Experiments at different conditions were conducted
to elucidate the sorption processes responsible for the observed solute retardation.

The results of this study agreed fairly well with results obtained from previous
batch equilibrium experiments. Retardation of anions was minimal under all conditions, consistent with the batch sorption experiments and with the reported minimal
anion exchange capacities of zeolites. Based on the flow interruption technique, it
was determined that nonequilibrium sorption was not a factor in Cr^{VI} sorption,
whereas Se^{IV} retardation was affected to a small degree by nonequilibrium sorption
at the higher flow rates. At the neutral or slightly basic pH of the experiments, very
little retardation was expected, but the TSM/TRM predicted slightly higher retardation for Se^{IV} compared to Cr^{VI}, consistent with batch sorption experiments. The
retardation factors estimated from batch and column experiments were in good
agreement.

Cation experiments were consistent with much higher solute retardation through
the zeolitized tuffs. For Pb and Sr, although a complete breakthrough was not
obtained at the lower ionic strength, the absence of breakthrough was consistent
with expected retardation factors in excess of 2000, as predicted by batch equilibrium

sorption experiments, even at high average linear velocities. At higher ionic strengths, however, where, based on batch equilibrium experiments, retardation was expected to be lower by at least an order of magnitude, the obtained results agreed again fairly well with these equilibrium experiments. Retardation factors obtained from batch and column experiments did not differ by more than a factor of two and, typically, the agreement was much better. A limited discrepancy between retardation factors derived from batch and column experiments could be explained because of the particle sizes and, therefore, slightly different specific surface areas, used in the two different sets of experiments. At the high flow rates employed during these experiments, sorption nonequilibrium was observed and was confirmed by the flow interruption technique. The estimated retardation factors, however, were similar, regardless of the model used (equilibrium vs. two-site/two-region).

ACKNOWLEDGMENTS

Financial support for this project was provided by the U.S. Department of Energy, Nevada Operations Office, under Contract DE-AC08-95NV11508. The continued support of Robert Bangerter is greatly appreciated. We thank C.J. Bruton and J. Wille for their careful review of the manuscript. Reference herein to any specific commercial product, process, or service by trade name, trademark, manufacturer, or otherwise, does not necessarily constitute or imply its endorsement, recommendation, or favoring by the United States Government or any agency thereof or its contractors or subcontractors.

REFERENCES

Baes, C.F., Jr. and R.E. Mesmer (1986). *The Hydrolysis of Cations*. Robert E. Krieger, Malabar, FL, 489 pp.

Bahr, J.M. (1989). Analysis of nonequilibrium desorption of volatile organics during field test of aquifer decontamination, *J. Contam. Hydrol.* 4, 205-222.

Bailey, S.E., T.J. Olin, R.M. Bricka, and D. Adrian (1999). A review of potentially low-cost sorbents for heavy metals, *Wat. Res.* 33, 2469-2479.

Ball, W.P. and P.V. Roberts (1991). Long-term sorption of halogenated organic chemicals by aquifer material. 2. Intraparticle diffusion, *Environ. Sci. Technol.* 25, 1237-1249.

Bargar, J.R., G.E. Brown, Jr., and G.A. Parks (1998). Surface complexation of Pb(II) at oxide-water interfaces: III. XAFS determination of Pb(II) and Pb(II)-chloro adsorption complexes on goethite and alumina, *Geochim. Cosmochim. Acta.* 62, 193-207.

Barrer, R.M. (1978). *Zeolites and Clay Minerals as Sorbents and Molecular Sieves*. Academic Press, New York, 497 pp.

Bernot, P.A. (1999). "Modeling the Rates of Lead and Strontium Uptake by Zeolitized Tuff from Rainier Mesa, Nevada Test Site, Nevada," M.S. Thesis, University of Nevada, Las Vegas.

Breck, D.W. (1974). *Zeolite Molecular Sieves: Structure, Chemistry, and Use*. John Wiley & Sons, New York, 771 pp.

Brunauer, S., P.H. Emmett, and E. Teller (1938). Adsorption of gases in multimolecular layers, *J. Am. Chem. Soc.* 60, 309-319.

Brusseau, M.L., R.E. Jessup, and P.S.C. Rao (1990). Sorption kinetics of organic chemicals: evaluation of gas-purge and miscible-displacement techniques, *Environ. Sci. Technol.* 24, 727-735.

Brusseau, M.L., R.E. Jessup, and P.S.C. Rao (1991). Nonequilibrium sorption of organic chemicals: Elucidation of rate-limiting processes, *Environ. Sci. Technol.* 25, 134-142.

Brusseau, M.L., P.S.C. Rao, R.E. Jessup, and J.M. Davidson (1989). Flow interruption: A method for investigating sorption nonequilibrium, *J. Contam. Hydrol.* 4, 223-240.

Bryant, E.A. and J. Fabryka-Martin (1991). Survey of hazardous materials used in nuclear testing, Report No. LA-12014-MS, Los Alamos National Laboratory, Los Alamos, NM.

Chisholm-Brause, C.J., K.F. Hayes, A.L. Roe, G.E. Brown, Jr., G.A. Parks, and J.O. Leckie (1990). Spectroscopic investigation of Pb(II) complexes at the γ-Al_2O_3/water interface, *Geochim. Cosmochim. Acta.* 54, 1897-1909.

Chisholm-Brause, C.J., A.L. Roe, K.F. Hayes, G.E. Brown, Jr., G.A. Parks, and J.O. Leckie (1989). XANES and EXAFS study of aqueous Pb(II) adsorbed on oxide surfaces, *Physica B.* 158, 674-675.

Cunningham, J.A., C.J. Werth, M. Reinhard, and P.V. Roberts (1997). Effects of grain-scale mass transfer on the transport of volatile organics through sediments. 1. Model development, *Water Resour. Res.* 33, 2713-2726.

Curtis, G.P., P.V. Roberts, and M. Reinhard (1986). A natural gradient experiment on solute transport in a sand aquifer 4. Sorption of organic solutes and its influence on mobility, *Water Resour. Res.* 22, 2059-2067.

Davis, J.A. and K.F. Hayes (1986). "Geochemical processes at mineral surfaces: an overview." In *Geochemical Processes at Mineral Surfaces*, (edited by J.A. Davis and K.F. Hayes), ACS Symposium Series, No. 323. American Chemical Society, Washington, D.C., 2-18.

Dyer, A. (1995). Zeolite surfaces and reactivity. In *Mineral Surfaces*, (Edited by D.J. Vaughan and R.A.D. Pattrick), Mineralogical Society Series, No. 5. Chapman & Hall, London, 333-354.

Dyer, A., A.S.A. Gawad, M. Mikhail, H. Enamy, and M. Afshang (1991). The natural zeolite, laumonite, as a potential material for the treatment of aqueous nuclear wastes, *J. Radioanal. Nucl. Chem.* 154, 265-276.

Fawaris, B.H. and K.J. Johanson (1995). Sorption of ^{137}Cs from undisturbed forest soil in a zeolite trap, *Sci. Total Environ.* 172, 251-256.

Francis, B.M. (1994). *Toxic Substances in the Environment. Environ. Sci. Technol.* Series. John Wiley & Sons, New York, 360 pp.

Freeze, R.A. and J.A. Cherry (1979). *Groundwater.* Prentice-Hall, Englewood Cliffs, NJ, 604 pp.

Fuller, C.C., J.A. Davis, G.A. Waychunas, and B.A. Rea (1993). Surface chemistry of ferrihydrite: Part 2. Kinetics of arsenate adsorption and coprecipitation, *Geochim. Cosmochim. Acta.* 57, 2271-2282.

Gaber, H.M., W.P. Inskeep, S.D. Comfort, and J.M. Wraith (1995). Nonequilibrium transport of atrazine through large intact soil cores, *Soil Sci. Soc. Am. J.* 59, 60-67.

Goltz, M.N. and P.V. Roberts (1988). Simulations of physical nonequilibrium solute transport models: application to a large-scale field experiment, *J. Contam. Hydrol.* 3, 37-63.

Gregg, S.J. and K.S.W. Sing (1982). *Adsorption, Surface Area and Porosity.* Academic Press, London, 303 pp.

Haggerty, G.M. and R.S. Bowman (1994). Sorption of chromate and other inorganic anions by organo-zeolite, *Environ. Sci. Technol.* 28, 452-458.

Hayes, K.F. and J.O. Leckie (1986). Mechanism of lead ion adsorption at the goethite-water interface. In *Geochemical Processes at Mineral Surfaces*, (edited by J.A. Davis and K.F. Hayes), ACS Symposium Series, No. 323. American Chemical Society, Washington, D.C., 114-141.

Hayes, K.F. and J.O. Leckie (1987). Modeling ionic strength effects on cation adsorption at hydrous oxide/solution interfaces, *J. Colloid Interface Sci.* 115, 564.

Hayes, K.F., C. Papelis, and J.O. Leckie (1988). Modeling ionic strength effects on anion adsorption at hydrous oxide/solution interfaces, *J. Colloid Interface Sci.* 125, 717-726.

Jacobs, P.H. and U. Forstner (1999). Concept of subaqueous capping of contaminated sediments with active barrier systems (ABS) using natural and modified zeolites, *Wat. Res.* 33, 2083-2087.

Katz, L.E., D.N. Humphrey, P.T. Jankauskas, and F.A. DeMascio (1996). Engineered soils for low-level radioactive waste disposal facilities: effects of additives on the adsorptive behavior and hydraulic conductivity of natural soils, *Hazardous Waste and Hazardous Materials.* 13, 283.

Klein, C. and C.S. Hurlbut, Jr. (1993). *Manual of Mineralogy (after J.D. Dana).* John Wiley & Sons, New York, 681 pp.

Li, Z. and R.S. Bowman (1997). Counterion effects of the sorption of cationic surfactant and chromate on natural clinoptilolite, *Environ. Sci. Technol.* 31, 2407-2412.

Loizidou, M. and R.P. Townsend (1987). Ion-exchange properties of natural clinoptilolite, ferrierite and mordenite: Part 2. Lead–sodium and lead-ammonium equilibria, *Zeolites.* 7, 153-159.

Ma, L. and H.M. Selim (1994). Tortuosity, mean residence time, and deformation of tritium breakthroughs from soil columns, *Soil Sci. Soc. Am. J.* 58, 1076-1085.

Ma, L. and H.M. Selim (1997). Evaluation of nonequilibrium models for predicting atrazine transport in soils, *Soil Sci. Soc. Am. J.* 61, 1299-1307.

Mackay, D.M., D.L. Freyberg, and P.V. Roberts (1986). A natural gradient experiment on solute transport in a sand aquifer 1. Approach and overview of plume movement, *Water Resour. Res.* 22, 2017-2029.

Malliou, E., M. Loizidou, and N. Spyrellis (1994). Uptake of lead and cadmium by clinoptilolite, *Sci. Total Environ.* 149, 139-144.

Maraqa, M.A., X. Zhao, R.B. Wallace, and T.C. Voice (1998). Retardation coefficients of nonionic organic compounds determined by batch and column techniques, *Soil Sci. Soc. Am. J.* 62, 142-152.

Ming, D.W. and F.A. Mumpton (1989). Zeolites in soils. In *Minerals in Soil Environments,* (edited by J.B. Dixon and S.B. Weed), Soil Science Society of America Book Series, Soil Science Society of America, Madison, WI, 873-911.

Oliviera, I.B., A.H. Demond, and A. Salehzadeh (1996). Packing of sands for the production of homogeneous porous media, *Soil Sci. Soc. Am. J.* 60, 49-53.

Papelis, C. and K.F. Hayes (1996). Distinguishing between interlayer and external sorption sites of clay minerals using X-ray absorption spectroscopy, *Colloids & Surfaces, A.* 107, 89-96.

Papelis, C., P.V. Roberts, and J.O. Leckie (1995). Modeling the rate of cadmium and selenite adsorption on micro- and mesoporous transition aluminas, *Environ. Sci. Technol.* 29, 1099-1108.

Parker, J.C. and M.T. van Genuchten (1984). Determining transport parameters from laboratory and field tracer experiments, Bull. 84-3, Virg. Agric. Exp. Stn.

Reedy, O.C., P.M. Jardine, G.V. Wilson, and H.M. Selim (1996). Quantifying the diffusive mass transfer of nonreactive solutes in columns of fractured saprolite using flow interruption, *Soil Sci. Soc. Am. J.* 60, 1376-1384.

Roberts, P.V., M.N. Goltz, and D.M. Mackay (1986). A natural gradient experiment on solute transport in a sand aquifer 3. Retardation estimates and mass balances for organic solutes, *Water Resour. Res.* 22, 2047-2058.

Roe, A.L., K.F. Hayes, C.J. Chisholm-Brause, G.E. Brown, Jr., G.A. Parks, K.O. Hodgson et al. (1991). *In situ* X-ray absorption study of lead ion surface complexes at the goethite-water interface, *Langmuir.* 7, 367-373.

Sax, N.I. (1981). *Cancer Causing Chemicals.* Van Nostrand Reinhold, New York, 466 pp.

Selim, H.M. and L. Ma (1995). Transport of reactive solutes in soils: a modified two-region approach, *Soil Sci. Soc. Am. J.* 59, 75-82.

Sloop, D.A. (1998). Equilibrium Studies of Ion Sorption on Zeolitized Tuff from Rainier Mesa, Nye County, Nevada, M.S. Thesis, University of Nevada, Las Vegas.

Stumm, W. (1992). *Chemistry of the Solid-Water Interface.* John Wiley & Sons, New York, 428 pp.

Weber, W.J., Jr. and F.A. DiGiano (1996). *Process Dynamics in Environmental Systems. Environ. Sci. Technol.* Series. John Wiley, New York, 943 pp.

Werth, C.J., J.A. Cunningham, P.V. Roberts, and M. Reinhard (1997). Effects of grain-scale mass transfer on the transport of volatile organics through sediments. 2. Column results, *Water Resour. Res.* 33, 2727-2740.

Wood, W.W., T.F. Kraemer, and P.P. Hearn, Jr. (1990). Intragranular diffusion: An important mechanism influencing solute transport in clastic aquifers?, *Science.* 247, 1569-1572.

Zhang, P.C. and D.L. Sparks (1990a). Kinetics and mechanisms of sulfate adsorption/desorption on goethite using pressure-jump relaxation, *Soil Sci. Soc. Am. J.* 54, 1266-1273.

Zhang, P.C. and D.L. Sparks (1990b). Kinetics of selenate and selenite adsorption/desorption at the goethite/water interface, *Environ. Sci. Technol.* 24, 1848-1856.

CHAPTER **3**

Modeling Competitive Sorption and Release
of Heavy Metals in Soils

Ruben Kretzschmar and Andreas Voegelin

INTRODUCTION

The mobility and fate of heavy metals in soils and sediments is governed to a large extent by competitive adsorption and desorption reactions at solid-water interfaces of minerals and soil organic matter (Alloway, 1995; McBride, 1989; Sparks, 1995). In addition, slower processes such as diffusion into microporous solids and formation of heavy metal bearing precipitates may lead to significant changes in metal speciation and solubility with time (Axe and Anderson, 1997; Barrow et al., 1989; Elzinga and Sparks, 1999; Roberts et al., 1999; Scheidegger et al., 1998; Thompson et al., 2000). Soils are by nature very complex multicomponent systems, in which many different ionic species in solution and solid phases with different surface properties are present. Comprehensive modeling of heavy metal sorption and release in soils is therefore a challenging task and should, ideally, take into account competitive sorption, dissolution and surface precipitation, sorbent heterogeneity, and kinetics (Lee et al., 1996; Temminghoff et al., 1995; Voegelin et al., 2001; Elzinga and Sparks, 1999; Paulson and Balistrieri, 1999; Roberts et al., 1999; Thompson et al., 2000).

In this chapter, we discuss current approaches of modeling heavy metal sorption in systems of increasing complexity, starting from simple model systems and progressing to soils contaminated with several heavy metals. First, we provide a brief overview of studies on competitive sorption of heavy metals to pure soil components, such as clay minerals, oxides, and humic substances. Then, our main focus will be the modeling of competitive sorption of heavy metals in soil materials. We will discuss various modeling approaches used in the past, and present a combined cation exchange/specific sorption model (CESS) in greater detail (Voegelin et al., 2001), including the extension of this model to describe Cd sorption in a wide range of

1-56670-531-2/01/$0.00+$1.50
© 2001 by CRC Press LLC

soil materials. Finally, a case study on Ca induced release of Cd, Zn, and Pb from smelter-contaminated soils will be presented.

COMPETITIVE SORPTION OF HEAVY METALS TO INORGANIC SOIL COMPONENTS

The sorption of heavy metals to clay minerals (Angove et al., 1998; Baeyens and Bradbury, 1997; Bradbury and Baeyens, 1997; Elzinga and Sparks, 1999; Puls et al., 1991; Scheidegger et al., 1996) and oxide minerals (Benjamin and Leckie, 1981b; Bruemmer et al., 1988; Christl and Kretzschmar, 1999; Gerth et al., 1993; Hayes and Leckie, 1986, 1987; Hoins et al., 1993; Lützenkirchen, 1997; Robertson and Leckie, 1998; Venema et al., 1996) has been the subject of numerous investigations. In most studies, solution pH was used as a master variable, since the adsorption of metal cations to oxide and clay mineral surfaces is strongly pH dependent (Sparks, 1995). The ionic strength dependence of metal cation sorption depends on the predominant sorption mechanisms. Weak ionic strength dependence of metal sorption has often been explained by specific inner-sphere complexation, while pronounced ionic strength dependence has been ascribed to nonspecific outer-sphere complexation mechanisms (Hayes and Leckie, 1986, 1987).

Many different variations of surface complexation models have been proposed to describe metal sorption to oxide mineral surfaces as a function of pH and ionic strength, including the constant capacitance model (Schindler and Gamsjäger, 1972; Schindler and Kamber, 1968), diffuse layer model (Huang and Stumm, 1973; Stumm et al., 1970), site-binding model (Yates et al., 1974), triple layer model (Davis et al., 1978; Davis and Leckie, 1978), variable surface charge – variable surface potential model (Bowden et al., 1973; Bowden et al., 1977), 1-pK basic Stern model (van Riemsdijk et al., 1986; Westall, 1986), and the generalized two-layer model (Dzombak and Morel, 1990). More recently, multisite approaches have been developed to account for the chemical heterogeneity of oxide surfaces (Hiemstra et al., 1989a, b). All surface complexation models are based on common principles: (1) sorption of ions takes place at specific surface sites; (2) sorption reactions are described by mass law equations; (3) surface charge results from sorption of ions; and (4) electrostatic effects on ion binding are taken into account (Dzombak and Morel, 1990). The main differences between the models are the formulation of surface protonation reactions (1-pK or 2-pK approach), the placement of ions at various distances within the electric double layer (Goldberg, 1992; Hiemstra and van Riemsdijk, 1996; Westall, 1986), and their ability to explicitly account for surface heterogeneity (Hiemstra et al., 1989b). A comparison of the performance of the most commonly used surface complexation models to describe metal sorption to oxides is given by Venema et al. (1996). Recently, several authors have developed theoretical approaches to predict surface complexation model parameters for different oxide minerals (Rustad et al., 1996; Sverjensky and Sahai, 1996).

Modeling proton and metal sorption to clay minerals is already more complex than for oxides, since clay minerals possess both pH independent negative charge on basal planes (due to isomorphic substitution) and pH dependent charge on edge surfaces and defects (due to surface protonation). Thus, combinations of permanently charged cation exchange sites and variably charged surface complexation sites have been used to describe proton and metal cation sorption to clay minerals, for example, kaolinite and montmorillonite (Bradbury and Baeyens, 1997; Kraepiel et al., 1998, 1999; Schindler et al., 1987).

The vast majority of studies on metal cation sorption to pure mineral components have been conducted using a so-called *indifferent* background electrolyte to control ionic strength, such as $NaNO_3$ or $NaClO_4$. It is usually assumed that the cations and anions of the background electrolyte only form weak outer-sphere complexes at the mineral surface by electrostatic attraction, depending on surface charge. Thus, sorption competition of indifferent cations, such as Na^+, with strongly sorbing heavy metals forming inner-sphere surface complexes is weak. From such studies, little information on possible competitive effects in heavy metal sorption to mineral surfaces can be obtained.

Few studies have been conducted to investigate the competitive sorption of two or more strongly sorbing metal ions to oxide minerals (Benjamin and Leckie, 1981a; Christl and Kretzschmar, 1999; Cowan et al., 1991; Palmqvist et al., 1999; Zasoski and Burau, 1988). The competitive sorption of several bivalent heavy metal cations (Cd, Cu, Zn, Pb) to amorphous iron oxyhydroxide (HFO) was studied by Benjamin and Leckie (1981a). The nearly complete lack of competition led them to hypothesize that different binding sites may be responsible for preferred sorption of the various metals. However, since HFO has an enormous surface area and metal sorption capacity, the observed lack of competition may also be related to low surface coverage and/or precipitation effects (Karthikeyan et al., 1999). Cowan et al. (1991) investigated Cd sorption to HFO in the presence of larger concentrations of alkaline-earth cations (Mg, Ca, Sr, Ba). Significant competition between Cd and Ca was observed, resulting in decreased Cd binding with increasing Ca concentration. Zasoski and Burau (1988) reported that Cd and Zn strongly compete with each other for sorption sites on MnO_2. Palmqvist et al. (1999) found clear competitive effects between Cu, Pb, and Zn on goethite, however, only at high surface loadings of the goethite with adsorbed metal cations. They predicted competitive effects using a constant capacitance surface complexation model based on acid-base titration and single-metal sorption data. Similarly, Christl and Kretzschmar (1999) investigated the competitive sorption of Cu and Pb to colloidal hematite particles. They showed that the competitive effects between both metals can be correctly predicted using a basic Stern surface complexation model calibrated on single metal sorption and acid-base titration data, provided that a surface site density between 5 and 10 sites/nm[2] is used in the model for hematite. Smaller site densities resulted in overestimation of competitive effects between Cu and Pb, while higher site densities resulted in underestimation of these effects.

COMPETITIVE SORPTION OF HEAVY METALS TO ORGANIC SOIL COMPONENTS

Several studies have been conducted on the competitive sorption of protons, heavy metal cations, and major cations to natural organic matter from soils and aquatic environments (Christensen et al., 1998; Hering and Morel, 1988; Kinniburgh et al., 1996, 1999; Mandal et al., 2000; Milne et al., 1995; Pinheiro et al., 1999; Tipping, 1993; Tipping et al., 1995; Town and Powell, 1993). In most studies, purified humic or fulvic acids were used as representative fractions of natural organic matter in soils. Kinniburgh et al. (1996) reported that the presence of Cd significantly reduced Ca binding to a purified peat humic acid. On the other hand, Cd binding was decreased by increasing Ca concentrations in the humic acid suspensions. The results of these studies show that bivalent cations, such as Ca and Mg, effectively compete with Cd for binding sites on humic acids. In contrast, the effect of Ca on Cu binding to purified peat humic acid containing a large amount of phenolic type functional groups was shown to be minimal, probably due to the much higher affinity of Cu for phenolic groups as compared to Ca (Kinniburgh et al., 1996). Similar lack of competition between Ca and Cu was reported by Hering and Morel (1988) in a study with Suwannee River humic acid. In contrast, McKnight and Wershaw (1994) found clear Ca competition for Cu binding to Suwannee River fulvic acid, indicating that there may be differences depending on the functional group composition of the humic substances. Pinheiro et al. (1999) studied sorption competition between Ca and Pb to a podsol fulvic acid and observed a clear reduction in Ca binding in the presence of Pb. Mandal et al. (2000) observed clear competitive effects of Ca and Mg on Ni binding to a soil fulvic acid.

Several different modeling approaches have been developed to comprehensively describe competitive ion binding to natural organic matter (Benedetti et al., 1995; Kinniburgh et al., 1996, 1999; Tipping, 1993, 1998; Tipping et al., 1995; Tipping and Hurley, 1992). Currently, two of the most successful models include the "consistent non-ideal competitive adsorption" (NICA) isotherm equation combined with the Donnan electrostatic model (Kinniburgh et al., 1999) and Tipping's Model VI (Tipping, 1998), respectively. The NICA-Donnan model is based on bimodal, quasi-Gaussian continuous affinity distributions, while Model VI is based on a bimodal but discretized affinity distribution. Both models account for the chemical heterogeneity of humic substances, variable stoichiometry of binding (e.g., monodentate, bidentate), competition between specifically bound ions, and electrostatic effects which influence nonspecific binding of counterions. Both models are flexible, but have a large number of adjustable parameters, which have to be determined by fitting experimental data. For a detailed discussion of modeling ion binding to humic substances, see Kinniburgh et al. (1998).

COMPETITIVE SORPTION OF HEAVY METALS TO SOIL MATERIALS

Applications of surface complexation models and ion binding models for humic substances to soil or sediment materials has been attempted by several authors with variable success (Benedetti et al., 1996; Davis et al., 1998; Goldberg, 1992, 1999; Papini et al., 1999; Temminghoff et al., 1997; Wang et al., 1997a; Wen et al., 1998; Zachara et al., 1992). Two approaches can be distinguished when applying surface complexation models to complex natural materials such as soils or sediments (Davis et al., 1998): first, the component additivity approach, in which metal sorption to the soil or sediment material is *predicted* using surface complexation constants obtained for relevant pure mineral components; second, the generalized composite approach, in which metal sorption is *described* by fitting the surface complexation constants to experimental data, assuming that the soil material behaves like a homogeneous composite material.

Major problems arise due to the complexity of soils and sediments, particularly with the first approach (Davis et al., 1998). For example, determination of all relevant mineral phases with their corresponding surface areas in contact with the liquid phase is, in most cases, a nearly impossible task. In addition, the composition of the equilibrium solution is not as easily controlled in experiments with soils as compared to single mineral phases bathed in simple electrolyte solutions. Application of surface complexation constants from pure mineral-electrolyte systems to complex natural materials is therefore connected with many uncertainties. Also, the surface charge and electrical surface potentials in soils are much more difficult to characterize than for simple oxide minerals, making the use of electrostatic models questionable. Acid-base titrations, which are commonly used for characterizing the charging behavior of oxides or humic substances as a function of pH and ionic strength, are much more difficult to interpret for soils because of additional proton-consuming reactions such as dissolution or precipitation of poorly crystalline phases. All these problems make it currently infeasible to reliably predict metal sorption with the component additivity approach, or to determine surface complexation parameters for soils, which have any physical meaning. Thus, a surface complexation model fitted to describe metal sorption in soils must be regarded as strictly empirical.

For practical engineering problems and risk assessments, it is currently preferable to use empirical models which should, however, fulfill the following five criteria: (1) they should be simple and have as few adjustable parameters as needed to comprehensively describe the experimental data; (2) they should be based on a consistent set of stoichiometrically balanced reaction equations which can easily be coupled to chemical speciation and reactive transport models; (3) they should be extendable to include additional components without refitting all parameters; (4) they should be predictive within the particular multicomponent system studied; and (5) they should be consistent with known major processes, such as cation exchange, specific adsorption, or precipitation reactions. In the following sections, we discuss different empirical modeling approaches used in the past to describe metal sorption to soil and sediment materials. We then present a reaction-based model (CESS) for Cd, Zn, and Ni sorption to acidic soil materials in greater detail, which fulfills all of the above criteria (Voegelin et al., 2001).

Empirical Isotherm Equations Describing Metal Sorption to Soil Materials

Sorption of heavy metals to pure mineral components is often investigated as a function of pH, resulting in the well-known adsorption edge plots exhibiting a steep increase in metal sorption with increasing pH (Sparks, 1995). With soil materials, however, it is much more difficult to vary the equilibrium pH over a wide range without creating problems due to dissolution of soil components or precipitation of heavy metals. Therefore, heavy metal sorption to soil materials is more often studied in terms of adsorption isotherms, in which metal sorption is measured as a function of solution concentration at fixed pH and background electrolyte concentration (Grolimund et al., 1995; Sauve et al., 2000a).

Various empirical isotherm equations have been used to describe metal sorption to soil materials. The most simple equation is the linear adsorption isotherm

$$q = K_d c \qquad (3.1)$$

where K_d is the so-called *distribution coefficient,* q is the amount of metal sorbed, and c is the metal concentration in solution. The K_d value reflects the solution volume that contains the same amount of the investigated element as one sorbent mass unit. When interpreting $\log K_d$ values, normalized K_d values (e.g., to L and kg) must be used. Anderson and Christensen (1988) investigated the adsorption of Cd, Zn, Ni, and Co to 38 different soil materials from Denmark at low adsorptive concentrations ($c < 0.15$ μM). Sorption isotherms determined with the batch technique were found to be linear and were described according to Equation 3.1. The $\log K_d$ values linearly increased with increasing solution pH. The resulting sorption equation for Cd is listed in Table 3.1. Note that in Table 3.1 we converted all equations reported in the original papers to the same reference units given in the footnote. Christensen et al. (1996) used a similar experimental setup to determine Cd and Ni sorption in 18 sandy aquifer materials from Denmark. Again, linear sorption isotherms were found, and the $\log K_d$ values were linearly correlated with solution pH. The resulting Cd sorption equation (Table 3.1) shows a similar pH dependence as reported by Anderson and Christensen (1988). However, the intercept of the pH-$\log K_d$ relationship was lower, reflecting the smaller total number of sorption sites in the aquifer material due to lower clay and organic matter contents. A similar equation was also presented by Sauve et al. (2000b) (Table 3.1), which was obtained from the extraction of contaminated soils. In a study on Cd sorption to 15 soil materials from New Jersey, Lee et al. (1996) found a closer correlation between $\log K_{om}$ and pH than between $\log K_d$ and pH (Table 3.1), with

$$K_{om} = K_d / om \qquad (3.2)$$

where *om* is the fractional soil organic matter content. When calculating K_{om} according to Equation 3.2, it is implicitly assumed that metal sorption occurs predominantly to the soil organic matter fraction.

Besides the linear sorption isotherm equation and the concept of the distribution coefficient, the well-known Freundlich isotherm and modifications thereof are often used to describe metal sorption to soils. In the Freundlich isotherm, the adsorbed amount is an exponential function of the concentration in solution

$$q = K_f c^{n_f} \qquad (3.3)$$

Table 3.1 Freundlich-Type Cd Sorption Equations

Data Source	Equation	Concentration Range	Electrolyte	NS	ND	r^2
Anderson and Christensen (1988)[a]	$\log(q_{Cd}) = -1.53 + 0.64pH + \log(c_{Cd})$	10^{-8} to 10^{-7} M	1 mM $CaCl_2$	38	117	0.78
Christensen et al. (1996)[a]	$\log(q_{Cd}) = -2.74 + 0.67pH + \log(c_{Cd})$	10^{-9} to 10^{-7} M	1 mM $CaCl_2$	18	16	0.94
Sauve et al. (2000b)[a]	$\log(q_{Cd}) = -0.49 + 0.59pH + \log(c_{Cd})$	10^{-9} to 10^{-6} M	10 mM KNO_3	64	64	0.73
Lee et al. (1996)[a]	$\log(q_{Cd}) = -0.55 + 0.46pH + \log(c_{Cd})$	10^{-6} to 10^{-4} M	10 mM $NaNO_3$	15	194	0.80
Lee et al. (1996)[a]	$\log(q_{Cd}) = 1.08 + \log(om) + 0.48pH + \log(c_{Cd})$	10^{-6} to 10^{-4} M	10 mM $NaNO_3$	15	194	0.93
Boekhold and van der Zee (1992)[a]	$\log(q_{Cd}) = -0.78 + \log(oc) + 0.50pH + 0.77\log(c_{Cd})$	10^{-6} to 10^{-4} M	10 mM $CaCl_2$	1	18	1.00
Buchter et al. (1989)[b]	$\log(q_{Cd}) = 2.24 + \log(oc) + (1.18{-}0.075pH)\log(c_{Cd})$	10^{-7} to 10^{-3} M	5 mM $Ca(NO_3)_2$	11	110	
Temminghoff et al. (1995)[b]	$\log(q_{Cd}) = -3.81 + 0.69pH + 0.85\log(a_{Ca}) - 0.34\log(a_{Ca})$	10^{-6} to 10^{-4} M	I = 0.03 M	1	461	
Elzinga et al. (1999)[a]	$\log(q_{Cd}) = -1.40 + 0.44pH + 0.85\log(a_{Cd}) - 0.30\log(a_{Ca}) + 0.66\log(CEC)$	10^{-7} to 10^{-2} M	0.1 – 10 mM $CaCl_2$	>100	1125	0.79
Voegelin et al. (2000a)[b]	$\log(q_{Cd}) = -0.71 + 0.86\log(a_{Cd}) - 0.52\log(a_{Ca})$	10^{-7} to 10^{-2} M	0.1 – 10 mM $CaCl_2$	1	48	0.99

[a] The originally published equations were converted to the following SI units: q_{Cd} in mol/kg (sorbed Cd), c_{Cd} in mol/L (total dissolved Cd concentration), a_{Cd} unitless (free Cd^{2+} activity), a_{Ca} unitless (free Ca^{2+} activity), CEC in mol/kg (cation exchange capacity), c_{Cd} in mol/L (total dissolved Cd concentration), a_{Cd} unitless (free Cd^{2+} activity), a_{Ca} unitless (free Ca^{2+} activity), om in kg/kg (fractional organic matter content), and oc in kg/kg (fractional organic carbon content).
[b] The published data were reanalyzed using the same units as listed above.

NS = number of soils, ND = number of data points.

with K_f being the Freundlich coefficient and n_f the Freundlich exponent. If the parameter n_f is unity, Equation 3.3 reduces to the linear sorption isotherm. However, in many studies, the parameter n_f is found to be less than unity, indicating nonlinear sorption behavior in the concentration range analyzed. It is important to note that the K_f depends nonlinearly on the concentration units used. Different K_f values can therefore only be compared after converting them to the same units (see Table 3.1). In a combined laboratory and field study on Cd sorption in a contaminated soil, Boekhold and van der Zee (1992) applied a modified Freundlich equation to describe Cd sorption as a function of the fractional organic carbon content (oc) and pH:

$$\log(q/q_r) = \log K_{foc} + \log(oc) + 0.5pH + n_f(c/c_r) \qquad (3.4)$$

where K_{foc} is the Freundlich coefficient based on the fractional organic carbon content and q_r and c_r are reference units (e.g., mol/kg and mol/L, respectively) used to obtain logarithms of dimensionless numbers. The equation derived from batch Cd sorption experiments (Table 3.1) could be used to predict the Cd concentration in $CaCl_2$-extracts of the contaminated field soil, based on soil pH, organic carbon content, and total Cd concentration. Buchter et al. (1989) used the Freundlich isotherm to describe the adsorption of the cations Co^{2+}, Ni^{2+}, Cu^{2+}, Zn^{2+}, Cd^{2+}, Hg^{2+}, and Pb^{2+} (and the oxyanions VO_3^-, CrO_4^{2-}, BO_3^{3-}, PO_4^{3-}, AsO_4^{3-}, SO_4^{2-}, SeO_4^{2-}) to 11 soil materials. They found strongest cation adsorption for Pb > Cu ≈ Hg and similar sorption behavior for Co, Ni, Zn, and Cd. The n_f values of Cd, Zn, Ni, and Co were reported to decrease with increasing pH (Buchter et al., 1989). When we converted the reported K_f values for Co, Ni, Zn, and Cd from ppm to mol/kg and mol/L units and subsequently normalized them to the soil organic carbon content, we found virtually the same K_{foc} values for all four metals. Also, the respective exponents n_f became very similar. The $\log K_{foc}$ values in different soils were not correlated with pH, but remained rather constant, e.g., for Cd at 2.24 ($\pm$ 0.42 standard deviation, 11 soils). In contrast, the respective exponents n_f decreased with increasing pH. For Cd, the linear relationship was $n_f = 1.18 - 0.075$ pH ($r^2 = 0.89$, n = 11). The Cd sorption equation resulting from reanalyzing the data presented by Buchter et al. (1989) is given in Table 3.1.

It is apparent from Table 3.1 that the sorption equations presented so far differ significantly from each other. This is partly due to the different types of soil or sediment materials used and to different Cd concentration ranges investigated. The sorption equations, however, also depend strongly on the type and concentration of the background electrolyte, since background cations compete with heavy metal cations for sorption sites. Temminghoff et al. (1995) therefore investigated Cd sorption to an acidic soil as a function of pH, Cd concentration, and type of background electrolyte. The data can be described by a Freundlich-type adsorption equation accounting not only for pH and Cd activity, but also for competitive adsorption of Ca (Table 3.1). A similar equation was presented by Elzinga et al. (1999) (Table 3.1). They compiled a large collection of published data on Cd, Zn, and Cu sorption, including more than 100 soils, and derived the respective general Freundlich-type isotherms by multiple linear regression analysis. The resulting Cd sorption isotherm equation (Table 3.1) accounts for Cd activity, Ca activity, pH, and cation exchange capacity (CEC). Soil organic matter was not considered; however, the CEC and pH probably indirectly accounted for differences in organic matter content.

In all of the above-mentioned studies, simple and robust Freundlich-like equations proved useful in describing heavy metal adsorption as a function of the heavy metal concentration in solution and other parameters such as background cation concentration, solution pH, soil organic matter content, and cation exchange capacity. In general, a similar sorption behavior was observed for Co, Ni, Cd, and Zn. Higher sorption was found at higher pH levels and higher soil organic matter content, roughly following the general relation (for Cd)

$$\log\left(q_{Cd}\right) \approx \log K_{om} + \log(om) + 0.5\,pH + n_f \log\left(a_{Cd}\right) \tag{3.5}$$

However, it is apparent from Table 3.1 that the presented Cd sorption equations strongly depend on the studied soil materials and the experimental conditions. The use of the empirical Freundlich-type equations is therefore limited to the estimation of concentrations or sorbed amounts within the concentration range and the type of soil material they were derived from. Furthermore, Freundlich-type isotherms are limited to low surface coverages of adsorbed metals far from the sorption maximum. Since the Freundlich-type equations are not derived from stoichiometrically balanced reaction (mass law) equations, it is not possible to combine the equations obtained from different datasets to describe more complex systems. For example, Freundlich-type equations separately derived for Cd-Ca and Zn-Ca datasets, respectively, cannot be combined to obtain a unique model predicting metal sorption in the ternary system containing Cd, Zn, and Ca.

Empirical Reaction-Based Models Describing Metal Sorption in Soils

Besides the rather few applications of surface complexation models to soil or sediment materials discussed earlier (Davis et al., 1998; Goldberg, 1992; Goldberg, 1999; Wang et al., 1997a; Wen et al., 1998), simpler reaction-based sorption models have been used by several authors. Abd-Elfattah and Wada (1981) and Wada and Abd-Elfattah (1979) investigated heavy metal sorption to different soil materials discussing their results in terms of cation exchange selectivity coefficients. These coefficients were found to decrease with increasing equivalent fractions of heavy metals sorbed to the exchanger, which was explained by specific adsorption to high-affinity sites at low concentration levels and additional nonspecific adsorption at higher concentration levels. As in the studies of Buchter et al. (1989), Pb and Cu were found to exhibit strongest adsorption, followed by Zn, and finally Cd and Co, which showed approximately the same exchange behavior. Cavallaro and McBride (1978) investigated Cu and Cd sorption to acidic and calcareous soil materials. For the calcareous soil materials, precipitation was assumed to dominate metal binding. In the acidic soil materials, however, Cu was found to adsorb more strongly than Cd and Ca. Competition with Ca was found for both heavy metals, indicating the importance of cation exchange mechanisms. Although in many studies adsorption to cation exchange sites was suppressed by rather high background electrolyte concentrations, this sorption mechanism likely plays an important role in natural soil environments and should therefore not be neglected.

Voegelin et al. (2001) recently presented a combined cation exchange/specific sorption model (CESS) to describe the competitive sorption and release of Cd, Zn, and Ni in the presence of different concentrations of Ca, Mg, and Na. In the CESS model, cation exchange reactions are represented by the Gaines-Thomas convention (Gaines and Thomas, 1953). For example, the exchange between Ca and Cd is written as

$$CaX_2 + Cd^{2+} \rightleftharpoons CdX_2 + Ca^{2+} \tag{3.6}$$

where X denotes exchanger sites with charge -1. The activities of sorbed species are assumed to correspond to the charge equivalent fractions y_M

$$y_M = \frac{z_M q_M}{CEC} \tag{3.7}$$

where z_M is the charge of cation M, q_M is the amount of cation M sorbed (in mol/kg), and CEC is the cation exchange capacity of the soil material (in mol_c/kg). For the binary Cd^{2+}/Ca^{2+} system, the CEC can then be expressed as

$$CEC = 2q_{CdX_2} + 2q_{CaX_2} \tag{3.8}$$

Combining these equations yields the mass law equation of the cation exchange reaction

$$K_{CdCa} = \frac{y_{Cd} a_{Ca}}{y_{Ca} a_{Cd}} \tag{3.9}$$

where K_{CdCa} is the Gaines-Thomas exchange coefficient and a_M are the activities of the free metal ions in solution. From the charge balance and the mass law equation, the competitive sorption isotherm for Cd can be derived:

$$q_{CdX_2} = \frac{1}{2} CEC \frac{K_{CdCa} a_{Cd}}{K_{CdCa} a_{Cd} + a_{Ca}} \tag{3.10}$$

Figure 3.1 shows Cd adsorption isotherms to an acidic soil material (Riedhof soil) from Switzerland over wide adsorptive concentration ranges in different $CaCl_2$ background electrolytes (Voegelin et al., 2001). Selected properties of the Riedhof soil are provided in Table 3.2. At high Cd and low Ca concentrations, the isotherms reach an adsorption maximum which approximately corresponds to the CEC of the soil material. At low Cd concentrations, the adsorption isotherms exhibit a slope of

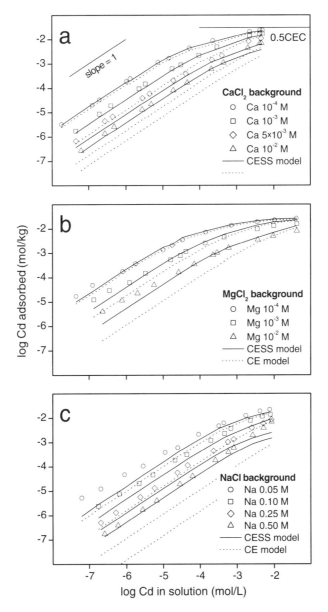

Figure 3.1   Cd adsorption to Riedhof soil in Ca, Mg, and Na backgrounds at pH 4.6. Maximum sorption is limited by the CEC. At low Cd concentrations, isotherms are linear. Sorption decreases with increasing background electrolyte concentration. Experimental data (symbols). Cation exchange (CE) model (dashed lines). Cation exchange/specific sorption (CESS) model calculations (solid lines). (From Voegelin, A., Vulava, V.M., and Kretzschmar, R., 2001. Reaction-based model describing competitive sorption and transport of Cd, Zn, and Ni in an acidic soil. *Env. Sci. Technol.* 35, 1651-1657. With permission.)

Table 3.2   Selected Properties of the Soils Riedhof, Hayhook, Evin-F, and Evin-T

	Land Use	Depth (cm)	Sieve Fraction[a] (mm)	pH	O.C. g/kg	Clay g/kg	CEC[d] (mmol/kg)
Riedhof	forest	15–25	0.063–0.400	4.6[b]	6.0	160	60
Hayhook	air base	25–50	<2.0	6.0[b]	1.1	100	45
Evin-F	forest	0–5	0.1–1.0	5.1[c]	52.0	230	200
Evin-T	arable crops	0–35	0.1–1.0	6.9[c]	17.0	210	188

[a] sieved aggregate size fraction
[b] in sorption experiments
[c] in 0.01 M CaCl$_2$ (2.5 mL/g)
[d] from column cation replacement experiment

unity on a log-log plot, indicating linear adsorption behavior of Cd. Finally, at any given Cd concentration, the amount of Cd adsorbed decreases with increasing Ca background concentration. All of these findings are in qualitative agreement with nonspecific adsorption by cation exchange. Thus, the cation exchange coefficient K_{CdCa} was adjusted to describe Cd adsorption at the lowest Ca background concentration, where cation exchange is most likely to dominate Cd sorption (Voegelin et al., 2001).

Figure 3.1a illustrates that the pure cation exchange model (Equations 3.6 to 3.10) strongly underestimates Cd sorption at higher Ca concentrations. Voegelin et al. (2001) explained this discrepancy by additional specific adsorption of Cd to a small number of binding sites exhibiting high Cd affinity. These high-affinity sorption sites would most closely correspond to the sorption sites investigated in most studies cited in the previous chapter, in which nonspecific sorption-to-cation exchange sites was suppressed by the high background electrolyte concentrations. Based on the results of these studies, the high-affinity sites are assumed to be related to the soil organic matter fraction. In the CESS model (Voegelin et al., 2001), sorption to such sites is represented by a set of competitive reaction equations:

$$Cd^{2+} + L^- \rightleftharpoons CdL^+ \tag{3.11}$$

$$Ca^{2+} + L^- \rightleftharpoons CaL^+ \tag{3.12}$$

where L^- denotes free sites with an assumed charge of -1, and ML^+ the sites occupied by the metal M. Assuming that the activities of sorbed species equal their concentrations, the resulting mass law equations are

$$K_{Cd} = \frac{\left(q_{CdL^+}/q_r\right)}{a_{Cd^{2+}}\left(q_{L^-}/q_r\right)} \tag{3.13}$$

$$K_{Ca} = \frac{\left(q_{CaL^+}/q_r\right)}{a_{Ca^{2+}}\left(q_{L^-}/q_r\right)} \tag{3.14}$$

Table 3.3   Cation Exchange/Specific Sorption Model Parameters for the Riedhof and Hayhook Soils

Riedhof			
Exchange site	CEC (mol/kg)	0.060	measured
	K_{CdCa}	0.5	adjusted (Figure 3.1a)
	K_{MgCa}	0.7	from (Voegelin et al., 2000b)
	K_{NaCa}	0.125	from (Voegelin et al., 2000b)
	K_{ZnCa}	0.5	estimated
Specific sorption site	L_T (mol/kg)	0.002	adjusted (Figure 3.1a)
	$K_{Cd,pH\ 4.6}$	21000	adjusted (Figure 3.1a)
	$K_{Ca,pH\ 4.6}$	450	adjusted (Figure 3.1a)
	$K_{Mg,pH\ 4.6}$	315	estimated
	$K_{Zn,pH\ 4.6}$	10000	adjusted (Figure 3.4a)
Hayhook			
Exchange Site	CEC (mol/kg)	0.045	calculated
	K_{CdCa}	1	estimated
	K_{KCa}	3.24	adjusted
	K_{NiCa}	1	estimated

where q_{CdL^+} and q_{CaL^+} are the respective sorbed species in mol/kg and q_r equals the reference unit of 1 mol/kg. The total site concentration L_T is given by

$$L_T = q_{L^-} + q_{CaL^+} + q_{CdL^+} \tag{3.15}$$

Solving these equations for q_{CdL^+} yields the respective competitive sorption Langmuir isotherm corresponding to the contribution of high-affinity sites

$$q_{CdL^+} = L_T \frac{K_{Cd}a_{Cd}}{1 + K_{Ca}a_{Ca} + K_{Cd}a_{Cd}} \tag{3.16}$$

The total amount of Cd adsorbed to the soil material, q_{Cd}, is calculated as the sum of Cd adsorbed to cation exchange sites, q_{CdX2}, and to specific adsorption sites, q_{CdL}. The parameters L_T, K_{Cd}, and K_{Ca} were adjusted to explain the discrepancies between the experimental data and the pure cation exchange model (Figure 3.1a). The combined exchange/specific sorption model (CESS) adequately describes the entire dataset spanning very wide ranges in Cd and Ca concentrations, with model parameters given in Table 3.3. The sorption coefficients for the specific sorption sites reflect the high preference for Cd over Ca at the specific sorption sites.

One of the five criteria which a competitive sorption model should meet is that it should be extendable to include additional components without refitting all model parameters. To test this, cation exchange coefficients for Ca-Mg and Ca-Na exchange in Riedhof soil were taken from a previous study (Voegelin et al., 2001) and combined with the Ca-Cd model to predict Cd sorption in the presence of various concentrations of Mg and Na, respectively. The only unknown model parameters were the sorption coefficients for Mg and Na to the competitive Langmuir sites. Since these sites are interpreted as being related mainly to the organic matter fraction, specific Na adsorption to these sites was assumed to be negligible ($K_{Na} = 0$), an assumption which is often made in ion binding models to humic substances (Kinniburgh et al., 1996). Further, the specific sorption sites were assumed to exhibit the same Ca-Mg selectivity as the cation exchange sites, that is, $K_{Mg} = K_{Ca}K_{MgCa}$ (Table 3.3). Figures 3.1b and 3.1c show experimental Cd sorption isotherms and corresponding model predictions with the CESS model extended for Mg and Na as additional components. Again, the pure cation exchange model (dashed lines) strongly underestimated Cd sorption at high Mg or Na concentrations. The CESS model (solid lines) yielded a good prediction of Cd sorption in Mg background electrolytes and also correctly predicted the trends of Cd sorption in Na background electrolyte solutions.

The last Freundlich-type Cd sorption equation presented in Table 3.1 was obtained from the Riedhof Cd sorption dataset in Ca background. The simple equation with three adjustable parameters also describes the Cd-Ca dataset accurately, as indicated by the r^2 value of 0.99. However, the Freundlich-type approach has several disadvantages, which can be illustrated nicely at this point. First, it would not be possible to extend the Freundlich equation to Cd sorption in Mg and Na background electrolyte solutions without refitting all model parameters. The stepwise development of a multicomponent model for complex soil systems is therefore not practical based on Freundlich-type equations. Second, the isotherm slope obtained from the Freundlich equation is 0.86 (Table 3.1), in agreement with the slopes reported by Elzinga et al. (1999) and Temminghoff et al. (1995). As discussed above, it is apparent from Figure 3.1a that the slope of the Cd isotherms is unity at low Cd concentrations as in the studies of Anderson and Christensen (1988) and Christensen et al. (1996). At higher Cd concentrations, the slopes of the Cd sorption isotherms decrease and approach zero when the sorption maximum is reached. The slope obtained in a Freundlich fit therefore depends on the experimental Cd concentration range used and the sorption maximum of the soil material. Thus, the obtained Freundlich parameters are not unique but depend on the experimental conditions chosen (Jenne, 1998).

Another of the five criteria a competitive sorption model should meet is that it can be incorporated into chemical speciation and reactive transport codes. To demonstrate this, the CESS sorption model was implemented with the speciation/transport code ECOSAT (Keizer et al., 1993) to predict Cd transport in soil column experiments (Voegelin et al., 2001). Figure 3.2 shows the influence of Ca and Cd concentration in the column influent on breakthrough of Cd in the column effluent. Riedhof soil material was packed into a column and prequilibrated with $CaCl_2$ background electrolyte at pH 4.6. At zero pore volumes, the influent was switched to a $CaCl_2$ solution also containing $CdCl_2$. After complete Cd breakthrough, the influent was switched back to the Cd-free $CaCl_2$ background electrolyte. Overall, excellent agreement between experiments and predictions was observed. As expected, the Cd concentration (10^{-5} or 10^{-6} M), which was within the linear range of the adsorption isotherms (Figure 3.1a), had little effect on Cd retention. In contrast, Cd breakthrough was strongly influenced by the Ca concentration in solution, which was correctly predicted by the model. Also the breakthrough curves corresponding to Cd desorption were correctly predicted, indicating that Cd sorption under the acidic conditions and the timescale of the experiments (minutes to weeks) was reversible.

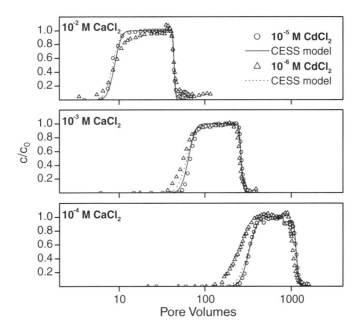

Figure 3.2 Cd transport in Riedhof soil at different $CaCl_2$ background electrolyte concentrations. After Cd breakthrough, columns were flushed with the $CaCl_2$ background electrolyte. Experimental data (symbols) and CESS model predictions (lines).

Predicting Sorption Competition Between Heavy Metals

Contaminated soils often contain elevated concentrations of several different heavy metals, which could potentially compete with each other for sorption to soil solid phases. Such sorption competition between heavy metals has been reported by several authors. In a study on competitive sorption of Cu, Co, and Ni to a Ca-saturated, coarse-loamy soil, Harter (1992) found that Cu sorbs more strongly than does Co and Ni. For Co and Ni, they observed desorption of Ca in a range similar to heavy metal adsorption, again indicating that cation exchange was a relevant adsorption process. Both Co and Cu were found to compete with Ni, however, the effect was more pronounced for Cu. Other authors report on competitive sorption of Cu, Ni, and Zn to a clayey kaolinitic soil (Bibak, 1997) or of Cs, Co, and Sr to a loamy soil (Gutierrez and Fuentes, 1991). These authors modeled their data with the Sheintuch-Rebhun-Sheindorf equation, which is also a Freundlich-type multi-component adsorption isotherm. Competitive sorption effects of Co, Ni, Zn, Pb, Cu, and Cr on Cd adsorption to sandy soil materials were reported by Christensen (1987a). For the competitive adsorption of Zn and Cd in Ca background, a model based on a competitive Langmuir equations was derived (Christensen, 1987b).

Wang et al. (1998) investigated competitive sorption effects between Cd and Ni in soil column experiments with a coarse-loamy soil (see Hayhook, Table 3.2). Soil material packed in a soil column was equilibrated with 10 mM KNO$_3$. At zero pore volumes, the influent was switched to one of the following solutions: (a) 1 mM Cd(NO$_3$)$_2$, (b) 1 mM Ni(NO$_3$)$_2$, or (c) 1 mM Cd(NO$_3$)$_2$ plus 1 mM Ni(NO$_3$)$_2$, all in 10 mM KNO$_3$ background electrolyte. After heavy metal breakthrough was complete, the influent solution was switched back to the pure background electrolyte. Similar transport behavior for Cd and Ni and increased mobility of both metals in the competitive experiment were observed. The authors modeled their breakthrough data using a linear sorption model, a nonlinear Freundlich model ($n_f = 0.334$ for Cd; $n_f = 0.373$ for Ni), and a nonequilibrium, nonlinear sorption model. The nonlinear Freundlich model yielded a much better prediction than a linear sorption model did, but additional improvements by accounting for nonequilibrium effects were minimal. Also in this study, the major disadvantages of Freundlich-type sorption models for multicomponent systems are apparent. First, the Freundlich model failed to predict the initial sharp decreases in heavy metal concentrations after the influent was switched back to heavy metal free solutions. Such fronts are often referred to as *normality fronts*, because they are related to the change in normality of the influent solution, which propagates through the column without retardation. Second, using the Freundlich equation, the authors were not able to predict heavy metal breakthrough in the experiment where both Cd and Ni were present and competed for binding sites. Again, the Freundlich equations could not be combined to define a unique model for the more complex system in which metal competition is important.

The lines in Figure 3.3 are predictions of the experimental breakthrough curves obtained with a pure cation exchange model using the Gaines-Thomas convention. Corresponding model parameters are provided in Table 3.3. Here, we assumed that the soil has the same cation exchange selectivity for Cd, Ni, and Ca. This was consistent with the batch adsorption and column breakthrough data presented by Wang et al. (1997b) and Wang et al. (1998). Due to the high heavy metal concentrations used by Wang et al. (1998), cation exchange was the major sorption mechanism. Thus, additional types of sites accounting for specific adsorption were not necessary in this case. Figure 3.3 shows that the simple cation exchange model not only yields a good description of Cd and Ni breakthrough in KNO$_3$ electrolyte background, but also correctly predicts the competitive experiment in which Cd and Ni were present in equal concentrations.

Competition between Cd and Zn was investigated in column transport experiments with the loamy Riedhof soil (Voegelin et al., 2001). Figure 3.4 shows breakthrough curves of Cd ($c_0 = 10^{-6}$ M CdCl$_2$) and Zn ($c_0 = 10^{-4}$ M ZnCl$_2$) in 10^{-3} M and 10^{-2} M CaCl$_2$ solutions, respectively. At the lower CaCl$_2$ concentration, Zn breakthrough occurred slightly earlier than Cd breakthrough (Figure 3.4a). While the cation exchange coefficient for Zn-Ca exchange was assumed to be the same as for Cd-Ca exchange, the high-affinity Zn sorption coefficient was adjusted to fit the Zn breakthrough curve shown in Figure 3.4a. With these additional parameters (Table 3.3), the model correctly described the coupled breakthrough patterns of Cd, Zn, and Ca in the presence of 10^{-3} M CaCl$_2$ background electrolyte in the influent. Interestingly, Cd breakthrough occurred slightly ahead of Zn breakthrough when the

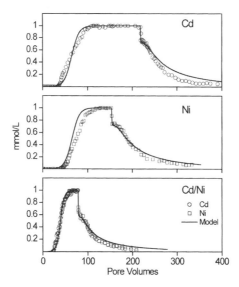

Figure 3.3   Cd, Ni, and combined Cd/Ni breakthrough in Hayhook soil. After breakthrough of Cd and/or Ni, the influent was switched back to the pure 10 mM KNO_3 background electrolyte. (Experimental data from Wang, W.-Z., Brusseau, M.L., and Artiola, J.F., 1998. Nonequilibrium and nonlinear sorption during transport of cadmium, nickel, and strontium through subsurface soils. In *Adsorption of Metals by Geomedia*. Jenne, E.A., Ed., Academic Press, San Diego; model calculations from Voegelin, A. and Kretzschmar, R., 2001. Competitive sorption and transport of Cd(II) and Ni(II) in soil columns: application of a cation exchange equilibrium model (in press). With permission.)

$CaCl_2$ concentration in the influent was increased to 10^{-2} M (Figure 3.4b). The CESS model correctly predicted this effect, which was due to the elevated Cl^- concentration. Chloride ions form stronger complexes with Cd than with Zn, thus preferentially increasing Cd mobility by the formation of $CdCl^+$ complexes in solution (Doner, 1978).

Toward a Generalized Sorption Model for Cd in Noncalcareous Soils

The cation exchange/specific sorption model (CESS) presented in the previous section proved capable of describing a large dataset of Cd sorption to the acidic Riedhof soil material at various $CaCl_2$, $MgCl_2$, and NaCl background electrolyte levels. In its present form, however, the model is conditional to the Riedhof soil at pH 4.6. It could also be calibrated for other soils at constant pH values. For many practical purposes, it would be highly desirable to develop models which can be used to estimate heavy metal sorption in a wide variety of soils from other soil properties, such as pH, organic matter content, and cation exchange capacity.

In order to obtain a generalized CESS model to describe Cd sorption to a wide variety of soils with different pH and organic matter content, a large collection of data on Cd sorption to soil materials in the presence of Ca background electrolytes was compiled from the literature. The generalized cation exchange/specific sorption model was based on several assumptions.

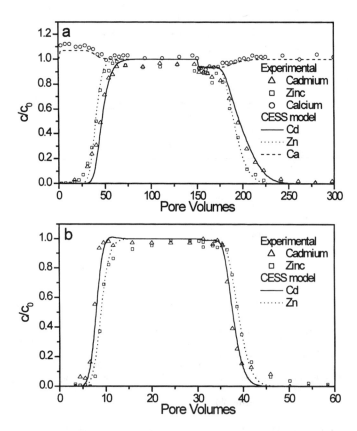

Figure 3.4   Coupled breakthrough of 10^{-6} M Cd and 10^{-4} M Zn in 10^{-3} (a) and 10^{-2} M (b) CaCl$_2$ background electrolyte in Riedhof soil at different Ca electrolyte concentrations. (Experimental data [symbols] and CESS model calculations [lines] from Voegelin, A., Vulava, V.M., and Kretzschmar, R., 2001. Reaction-based model describing competitive sorption and transport of Cd, Zn, and Ni in an acidic soil. *Env. Sci. Technol.* 35, 1651-1657. With permission.)

First, it was assumed that Cd sorption to cation exchange sites is pH-independent and nonspecific, i.e., K_{CdCa} was set to 1 (Equation 3.9; Table 3.4). Some pH dependence of the amount of Cd adsorbed to the exchanger, however, arises from the decrease of the measured effective CEC with decreasing pH (Curtin and Rostad, 1997). The second assumption was that the concentration of specific sorption sites is proportional to the soil organic matter content:

$$L_T = A \times oc \tag{3.17}$$

where A is the specific sorption site concentration in mol/kg organic carbon and oc the fractional soil organic carbon content in kg/kg. Finally, it was assumed, that the extent of sorption to the specific sorption sites increases with increasing pH. The second and third assumptions are in general agreement with results obtained in the previously

Table 3.4   Summary of the Generalized Cation Exchange/ Specific Sorption Model Describing Cd Sorption as a Function of pH, Organic Carbon Content, and Ca Concentration in a Wide Range of Soils

$$q_{Cd} = \frac{1}{2} CEC \frac{K_{CdCa} a_{Cd}}{a_{Ca} + K_{CdCa} a_{Cd}} + A \times oc \frac{K_{Cd,0} a_H^{-x_{Cd}} a_{Cd}}{1 + K_{Ca,0} a_H^{-x_{Ca}} a_{Ca} + K_{Cd,0} a_H^{-x_{Cd}} a_{Cd}}$$

Cation Exchange Site		
	CEC	measured (mol/kg)
	K_{CdCa}	1

Specific Sorption Site		
	oc	measured (kg/kg)
	A	0.154 mol/kg
	$\log K_{Cd,0}$	−3.77
	x_{Cd}	1.76
	$\log K_{Ca,0}$	−2.15
	x_{Ca}	1.04

cited studies, which revealed that specific sorption (i.e., metal sorption in the presence of high background electrolyte concentrations) roughly follows Equation 3.5 and therefore increases with increasing pH and organic matter fraction. To implement the pH dependence of the specific sorption reactions, the sorption coefficients K_{Cd} and K_{Ca} were assumed to increase with increasing pH according to:

$$\log K_{Cd,pH} = \log K_{Cd,0} + x_{Cd} pH \qquad (3.18)$$

$$\log K_{Ca,pH} = \log K_{Ca,0} + x_{Ca} pH \qquad (3.19)$$

$\log K_{Cd,pH}$ and $\log K_{Ca,pH}$ are the conditional sorption coefficients for reactions 3.11 and 3.12, respectively, at the experimental pH. x_{Cd} and x_{Ca} are adjustable parameters giving the pH dependence of the respective sorption coefficients. $\log K_{Cd,0}$ and $\log K_{Ca,0}$ are the sorption coefficients at pH 0. They were calculated from the conditional Riedhof model sorption coefficients for pH 4.6 (Table 3.3):

$$\log K_{Cd,0} = \log K_{Cd,pH4.6} - 4.6 x_{Cd} \qquad (3.20)$$

$$\log K_{Ca,0} = \log K_{Ca,pH4.6} - 4.6 x_{Ca} \qquad (3.21)$$

The assumption of pH-dependent conditional sorption coefficients is equivalent to the formulation of, for example, the Cd adsorption reaction with simultaneous proton release:

$$Cd(H_2O)_6^{2+} + L^- \quad LCd(H_2O)_{6-x}(OH)_x^{(2-x)^+} + x_{Cd} H^+ \qquad (3.22)$$

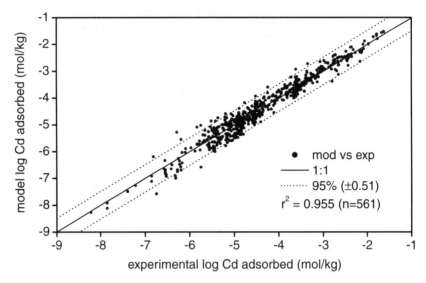

Figure 3.5   Comparison of modeled and experimental Cd adsorbed for a wide range of soil materials. Model calculations based on the generalized model presented in Table 3.4.

The constant sorption coefficient for this reaction is then equal to the above-defined $\log K_{Cd,0}$, and x_{Cd} and x_{Ca} can be interpreted as the fractional proton release due to Cd and Ca adsorption. Equations similar to Equation 3.22 are often cited in order to explain the fact that Cd sorption is proportional to a fractional power of the proton activity (Anderson and Christensen, 1988; Temminghoff et al., 1995; Hesterberg et al., 1993).

When the CESS model was extended to describe the literature dataset on Cd adsorption, only the three new model parameters x_{Cd}, x_{Ca}, and A were adjusted. All other parameters were either measured (pH, oc, CEC), estimated (K_{CdCa}), or derived from the conditional Riedhof model ($K_{Cd,0}$, $K_{Ca,0}$). The generalized model is summarized in Table 3.4. Figure 3.5 depicts the comparison of model predictions and experimental data for the entire literature dataset. The model is able to capture the general sorption behavior of Cd, notably in approximately 60 soils highly variable in pH (3.5 to 8.0), organic carbon content (0.01 to 20%), CEC (9 to 480 mmol/kg), and at Cd concentrations ranging from from 10^{-10} to 10^{-2} mol/L. The 95% confidence interval for model estimates is approximately ± half a log unit, i.e., ± a factor of 3 on a linear scale. Hence, the model can give a rather good estimate of the approximate range of the adsorbed Cd. With respect to the application of the model to transport modeling, it is, however, important to notice that in transport experiments, adsorbed amounts are linearly translated to the time scale, which is normally linearly plotted. An uncertainty of ± factor 3 in adsorbed amounts, therefore, has a huge impact on the shape and position of modeled fronts. Because transport modeling requires more accurate sorption parameters than the description of sorption data over several orders of magnitude, general parameters normally have to be further adjusted to a specific soil material to obtain an accurate description of transport patterns. Nevertheless, the presented generalized cation exchange/specific sorption model proves to be flexible enough not only to be applied to one specific soil, but also to describe the general sorption behavior of Cd to a wide variety of soil materials. In contrast to a simple Freundlich-type description, this model allows us to conceptually separate between nonspecific and specific adsorption reactions. In addition, it is extendable to further sorption reactions and competing components, as was demonstrated in the previous sections.

MODELING HEAVY METAL RELEASE FROM CONTAMINATED SOILS

Modeling heavy metal release and transport in contaminated soils requires that the equilibria and kinetics of desorption and/or relevant dissolution reactions is well understood. In the most simple case, heavy metals are reversibly adsorbed to the soil matrix and desorption kinetics is fast, such that the local equilibrium assumption can be applied in transport modeling. One example where such a simple modeling approach can be applied is the laboratory column experiment recently published by Wang et al. (1997b). Soil columns packed with the Hayhook soil (Table 3.2) were equilibrated with a solution containing 1 mM Cd(NO$_3$)$_2$ and 10 mM KNO$_3$. The adsorbed Cd was subsequently remobilized by leaching the columns with CaCl$_2$ solutions of different concentrations (1.25 to 50 mM). Due to Ca-Cd cation exchange, the CaCl$_2$ concentration in solution had a pronounced effect on the Cd elution pattern (Figure 3.6). The maximum height of the Cd mobilization peak increased with increasing Ca concentration in the influent, while the average Cd elution time decreased drastically. Note, that the Cd elution peak at 10 mM CaCl$_2$ shows a small shoulder preceding the peak maximum. At lower CaCl$_2$ concentrations, this feature appears as a precedent plateau in Cd concentration. These features are due to K in the system (from the 10 mM KNO$_3$), which is exchanged by Ca more readily than Cd and therefore produces an earlier cation exchange front.

Due to the rather high Cd concentration levels used by Wang et al. (1997b), the experiment can again be described with a simple cation exchange model (Equations 6-10). The lines in Figure 3.6 show model calculations based on the parameters listed in Table 3.3 (Voegelin and Kretzschmar, 2001). We used the same parameters as in the previous section for modeling the experiments shown in Figure 3.3. The cation exchange model correctly describes Cd release in this ternary (Cd-Ca-K) system, including the shoulder and plateaus in Cd concentration due to Ca-K exchange. In this case, a simple reaction-based multicomponent model can describe all features of heavy metal release.

From the Laboratory to the Field

The ultimate challenge is to model heavy metal release in field contaminated soils, where the contamination was caused by decade-long inputs of metal containing solid phases such as organic materials (e.g., sewage sludge), metal oxides, sulfides, or slag materials (e.g., smelter dust emission). Over time, heavy metals are released

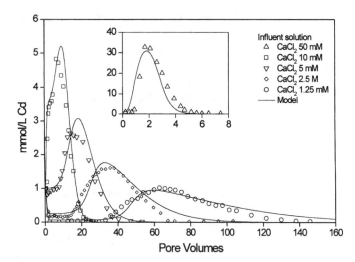

Figure 3.6 Cd elution from Hayhook soil at different Ca concentrations. Experimental data (symbols) and model calculations (lines). (Experimental data from Wang, W.-Z., Brusseau, M.L., Artiola, J.F., 1997b. The use of calcium to facilitate desorption and removal of cadmium and nickel in subsurface soils. *Journal of Contaminant Hydrology.* 25, 325-336; model calculations from Voegelin, A. and Kretzschmar, R., 2001. Competitive sorption and transport of Cd(II) and Ni(II) in soil columns: application of a cation exchange equilibrium model (in press). With permission.)

from these solids and can subsequently be readsorbed to soil-borne solid phases. Studies on model systems have shown that, depending on the prevailing chemical conditions, metals can also be immobilized by the formation of new solid phases such as hydroxides, hydrotalcites, or phyllosilicate structures (Ford et al., 1999; Scheidegger et al., 1998; Thompson et al., 2000; Xia et al., 1997). In general, acidic conditions, high organic matter content, high CEC, and low concentrations of major cations tend to favor the adsorption of heavy metals, while neutral to alkaline conditions favor the precipitation of new solid phases. Release of heavy metals from contaminated soils, therefore, also strongly depends on the chemical speciation of the contaminant and the soil conditions. The field situation is hence somewhat different from the experiments discussed so far, where heavy metals were added to soils in the form of dissolved heavy metal salts, and reaction times were in the range of minutes to weeks. Nevertheless, numerous results show that models and concepts derived from laboratory batch and column studies may be applicable to soils in the field (Boekhold and van der Zee, 1992; Sauve et al., 2000b; Hesterberg et al., 1993; Benedetti et al., 1996).

Case Study: Release of Cd, Zn, and Pb from Smelter-Contaminated Soils

In the following section, we present selected results of a recent study on heavy metal release from two contaminated soils in the vicinity of a large lead-zinc smelting plant (Pennaroya) in Evin-Malmaison, France (Kinniburgh et al., 2000; Morin et al.,

Table 3.5   Total, Batch BaCl$_2$-Exchangeable, and Column CaCl$_2$-Exchangeable Metal Contents of Soils Evin-F and Evin-T

	Totals (mmol/kg)[a]			Batch Exchangeable (mmol/kg)[b]						Column Exchangeable (mmol/kg)[c]		
	Zn	Cd	Pb	Zn	Cd	Pb	Ca	Mg	K	Zn	Cd	Pb
Evin-F	33.8	0.305	10.8	15.1	0.228	1.206	56.3	8.5	10.3	20.0	0.259	1.52
Evin-T	10.9	0.069	2.35	0.017	0.013	nd	82.6	3.1	6.2	0.0765	0.0311	nd

[a] HF-extract
[b] Unbuffered 0.1 M BaCl$_2$-extract (30mL/g) (Hendershot and Duquette, 1986)
[c] Cumulatively extracted in column experiments
nd = not detectable

1999). The smelter has been active since 1884 and has emitted heavy metal containing dusts until filters were installed in the late 1970s. This resulted in heavy contamination of surrounding soils with Zn, Pb, and Cd. The first soil, Evin-F, is a forested soil under poplar trees. The second soil, Evin-T, is in an arable field and has a plough layer of around 30-cm depth. The Evin-F soil is more acidic and has a higher organic matter content than the Evin-T soil. In both soils, the heavy metal concentrations are higher in the topsoil and decrease strongly with depth, indicative of anthropogenic metal pollution.

Selected soil properties of the Evin-F and Evin-T topsoils are given in Table 3.2. Heavy metal contents of the sieved aggregate fractions used in column leaching experiments are reported in Table 3.5. The ratio of Zn:Cd:Pb was similar in both soils, although Evin-F soil had higher total heavy metal contents than Evin-T soil (HF extract). This is due to the fact that the Evin-F topsoil was undisturbed, whereas Evin-T soil had a plow horizon within which the heavy metals were homogeneously distributed. Drastic differences were observed in the fraction of metals, which were exchangeable by 0.1 M BaCl$_2$ (Hendershot and Duquette, 1986). For example, 45% of the total Zn in Evin-F soil material was Ba exchangeable, while in Evin-T soil only 0.15% of the total Zn was exchangeable. The same trend is found for Cd and Pb, with Cd being more exchangeable and Pb being much less exchangeable than Zn. In both soils, Ca is the most abundant cation on the exchanger, with higher contributions of Mg and K in the Evin-F soil. Note that in Evin-F soil, the exchangeable Zn equivalent fraction is higher than that of Mg and K, and that Zn covers 15% (!) of the total exchange capacity. Recent XAFS results suggest that in the Evin-T soil a larger fraction of Zn is bound in hydrotalcite and phyllosilicate structures than in the Evin-F soil (Kinniburgh et al., 2000).

Figure 3.7 shows the results of a column leaching experiment with the Evin-F soil. A column was packed with soil material and eluted with solutions of increasing CaCl$_2$ concentration. Initially, the column was flushed with a 10^{-5} M CaCl$_2$ solution. During the initial 10 pore volumes, some colloidal soil particles and DOC were eluted from the column, resulting in slightly elevated heavy metal concentrations. Increasing the influent CaCl$_2$ concentration to 10^{-2} M at 26 pore volumes resulted in a pronounced elution pattern of other cations, including heavy metals. Note that the observed pattern is very similar for Cd and Zn, but distinctly different for Pb (Figure 3.7).

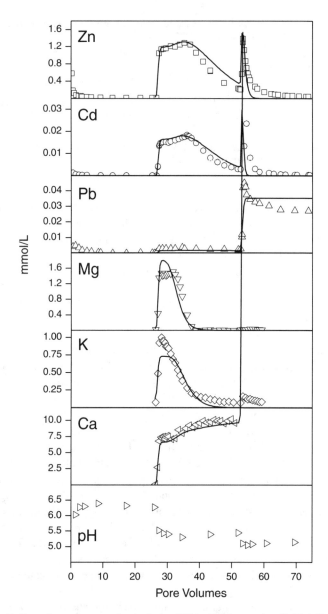

Figure 3.7 Cation release from contaminated soil Evin-F at increasing $CaCl_2$ influent concentration: 0 to 26 pore volumes 10^{-5} M $CaCl_2$, 26 to 52 pore volumes 10^{-2} M $CaCl_2$, >52 pore volumes 10^{-1} M $CaCl_2$. Due to the low pH, the high organic matter content, and the high mobility of heavy metals, the release can be described by a simple cation exchange model, and a similar elution behavior of Cd and Zn is observed. Experimental data (symbols) and model calculations (lines).

The lines in Figure 3.7 are preliminary model calculations based on a one-site cation exchange model accounting for sorption competition between Ca, Mg, K, Cd, and Zn. The total exchangeable metal pools and the cation exchange coefficients

Table 3.6   Model Parameters Used to Model Heavy Metal Leaching from a Smelter-Contaminated Soil (Evin-F)

	Exchangeable (mmol/kg)[a]	Equivalent Fraction (%)	K_{MCa}[b]
Zn	20.0	20.0	1.82
Cd	0.259	0.259	3.42
Pb	5.79	5.79	576
Mg	8.41	8.41	0.49
K	5.18	2.59	9.00
Ca	62.9	62.9	

[a] See text
[b] K_{MCa} = exchange coefficient for the exchange of Ca by the cation M

used in the model calculations are provided in Table 3.6. For Cd and Zn, the initial exchangeable amount was set to the cumulative extracted amount of the column experiment. For Pb, the initial exchangeable pool was taken from a column extraction experiment at 10^{-1} M $CaCl_2$ over more than 1000 pore volumes. Initial K and Mg were calculated from the column experiment, and the initial Ca was finally adjusted in order to account for the cation exchange capacity of the Evin-F soil (Table 3.2). Note that the amount of metals considered to be exchangeable apparently depends on the extraction method. Even though 0.01 M $CaCl_2$ is a weaker extract than 0.1 M $BaCl_2$, the column extracted metal amounts exceed those extracted in batch. While the batch is a closed system where equilibrium between solution and soil is reached, the soil in the column is permanently flushed and the metals are continuously removed from the system. The exchange coefficient K_{MgCa} given in Table 3.6 was calculated from a Mg/Ca column exchange experiment. The K_{KCa} exchange coefficient was calculated from measured pore water concentrations at the field site assuming equilibrium with the solid phase. The exchange coefficients for Cd, Zn, and Pb were finally adjusted so that the elution patterns could be accurately described. Although some deviations from the experimental data remain, the model captures all major features of the elution curves. The shoulder in the Cd and Zn elution peaks at approximately 26 to 34 pore volumes is related to the replacement of K and Mg by Ca. When Mg and K were depleted, Cd and Zn concentrations started to increase again. This trend overlapped with the depletion of the exchangeable Cd and Zn pools, which resulted in a sharp decrease in Cd and Zn concentrations after around 38 pore volumes. Increasing the $CaCl_2$ concentration to 10^{-1} M (at 50 pore volumes) resulted in a second sharp Cd and Zn peak and accelerated further depletion of the exchangeable pools. Note that the observed pattern is very similar to the elution pattern at 10 mM $CaCl_2$ in the experiments of Wang et al. (1997b) (Figure 3.6). Since $CaCl_2$ mobilizes mainly the adsorbed metal fractions, the similar elution patterns of Cd and Zn agree well with the similar sorption behavior of both metals in controlled Cd/Zn-transport experiments (Figure 3.4). In contrast, Pb showed a different elution pattern indicating much lower mobility. The observed pattern may either be due to sorption of Pb with high affinity or due to slow dissolution of Pb-bearing mineral phases. Note that a simple column extraction experiment as carried out with Evin-F soil yields much more information than the analogous batch extraction experiment. From the integration of the elution curves, the exchangeable amounts of heavy metal and major cations can be calculated. From the form of the elution curves, conditional exchange coefficients can be derived, giving a measure for the binding strength of different adsorbed cations. In addition, the elution pattern may also reveal the influence of kinetic effects on metal desorption depending on the flow rate applied in the experiment. Following the above elution experiment, the same column can be subsequently used for a Ca-Mg-Ca exchange experiment, for example, that allows us to determine the effective CEC of the soil material and the K_{MgCa} exchange coefficient. Finally, the leached soil material may be used to further investigate the nature of the residual heavy metal fraction.

In the calculations presented in Figure 3.7, we used a simple cation exchange model with only one type of sorption site. It would certainly be desirable to have a generalized model, such as the previously presented CESS model, which can be directly applied to predict heavy metal release from contaminated soils. However, several factors complicate the direct application of a generalized multisite model to a particular soil material. First, metal release and transport data are usually plotted on a linear time scale and, therefore, satisfactory transport predictions require very accurate sorption model parameters. For accurate transport calculations, the sorption model parameters need to be determined or estimated specifically for the respective soil material. The second problem is related to the initial distribution of adsorbed heavy metal cations to the different types of sorption sites when using equilibrium multisite models for transport calculations. In a simple one-site model, the amounts of exchangeable cations listed in Table 3.6 for Evin-F soil are simply assigned to the one type of sorption site. In contrast, when using a multisite model, the total adsorbed amounts have to be distributed over two or more types of sorption sites. This distribution may be based on further information, e.g., from parallel or sequential extractions or spectroscopic studies. If we assume that we know the correct distribution of adsorbed metals between the various types of sorption sites, the resulting initial state then ideally should represent the equilibrium with respect to the composition of the soil solution at the time of soil sampling. At this point, we can start a transport calculation. If we apply an equilibrium multisite model, then the transport code will immediately redistribute the adsorbed cations over all types of sites depending on the composition of the influent solution. In other words, the initial state of the sorbent phase is immediately changed, depending on the composition of the leaching solution applied. This however seems rather unlikely to occur in reality. The instantaneous redistribution of adsorbed metal cations can only be prevented by introducing slow kinetics of adsorption or desorption reactions into the model (Sparks, 1988). A simple one-site model has the advantage that redistribution of adsorbed metal cations is not possible. However, if chemically more realistic multisite models are to be applied to transport modeling, this example emphasizes the need for kinetic multisite approaches. Strawn and Sparks (2000) have recently shown that slow sorption and desorption processes of Pb in different soil materials are related to the soil organic matter content. Voegelin et al. (2001) found no significant kinetic effects or irreversibility with respect to the adsorption and desorption of Cd, Zn, and Ni in the acidic Riedhof soil material, which is low

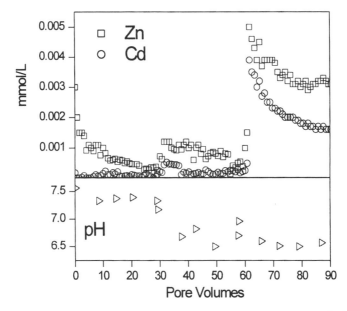

Figure 3.8   Heavy metal release from contaminated soil Evin-T at increasing $CaCl_2$ influent concentration: 0 to 26 pore volumes 10^{-5} M $CaCl_2$, 26 to 52 pore volumes 10^{-2} M $CaCl_2$, >52 pore volumes 10^{-1} M $CaCl_2$. At the pH of soil Evin-T, heavy metals are precipitated in rather than adsorbed to the soil material. Release cannot be modeled with a simple cation exchange model. Cd is considerably more mobile than Zn.

organic carbon content, suggesting that cation exchange reactions are rather fast (Table 3.2). In the CESS model for instance, this information might be incorporated by assuming that desorption from the specific sorption sites is slow, while adsorption and desorption reactions at the cation exchange sites are fast.

A similar metal release experiment with the Evin-T soil is presented in Figure 3.8. Here, a completely different elution pattern is observed. Zn and Cd concentrations in the column effluent were approximately 1000 and 100 times smaller than in the leaching experiment with Evin-F soil. A simple cation exchange model cannot describe the elution curves for Evin-T soil. It is likely that precipitation/dissolution reactions rather than sorption reactions determine Cd and Zn mobility in this soil material. This is consistent with the drastic differences found in exchangeable fractions of Zn and Cd in soils Evin-F and Evin-T, respectively (Table 3.5). In the experiment with Evin-T soil, Cd and Zn effluent concentrations did not reflect the total concentrations in the soil material, but Cd showed a significantly higher mobility than Zn. It is likely that the lower mobility of Zn is due to the lower solubility products of Zn oxides and hydroxides as compared to the respective Cd mineral phases. Pb was not found to elute significantly from Evin-T soil.

The results of this case study show, that an adjusted competitive sorption model based on one type of sorption site may describe the mobilization and release of heavy metals from a field-contaminated soil, provided that a large portion of the heavy metals is present in the readily exchangeable fraction. Competitive sorption is likely to control heavy metal release at low soil pH, high organic carbon content, and high cation exchange capacity. On the other hand, even in the case of precipitation/dissolution or kinetically controlled heavy metal release, an adequate multicomponent sorption model including kinetics will be necessary. Once heavy metals are released by dissolution, adsorption reactions at solid-water interfaces occur and will influence the subsequent fate of released metal cations and leaching into the groundwater.

CONCLUDING REMARKS

Predicting the fate of heavy metals in contaminated soils and the risk of metal mobilization and leaching into groundwater requires a quantitative understanding of the chemical behavior of heavy metals in complex multicomponent systems. Such knowledge can only be acquired based on a combination of macroscopic studies on metal behavior in multicomponent systems, microscopic and spectroscopic studies on metal sorption mechanisms at the molecular level, and the development of modeling tools. Macroscopic studies alone would remain highly empirical and would, in long term, not improve our understanding of relevant processes at the molecular level. Microscopic studies alone can provide new insights on sorption mechanisms, but would not directly improve our ability to quantitatively predict metal behavior in complex soil systems. The models to be developed should be simple, but they should consider the relevant chemical processes and comprehensively describe metal behavior in multicomponent soil systems under varying chemical conditions. A combined cation exchange/specific sorption model (CESS) was presented, which was used to describe Cd, Zn, and Ni behavior in acidic mineral soils. A first attempt was made to extend this model to describe Cd sorption to a wide range of soils as a function of Ca concentration, soil pH, and organic matter content. Such an extended CESS model might prove useful in risk assessment programs covering a wide range of soils. With respect to the modeling of heavy metal release from contaminated soils, the kinetics of desorption reactions needs further investigation.

ACKNOWLEDGMENTS

We are grateful to Kurt Barmettler for his technical support with performing transport experiments and to Ilona Heidmann for helpful comments on an earlier version of the manuscript. We also thank an anonymous reviewer for a very careful and constructive review.

REFERENCES

Abd-Elfattah, A. and Wada, K., 1981. Adsorption of lead, copper, zinc, cobalt, and cadmium by soils that differ in cation-exchange materials. *J. Soil Sci.* 32, 271-283.
Alloway, B.J. (Ed.) 1995. *Heavy Metals in Soils.* Chapman & Hall, London.
Anderson, P.R. and Christensen, T.H., 1988. Distribution coefficients of Cd, Co, Ni, and Zn in soils. *J. Soil Sci.* 39, 15-22.

Angove, M.J., Johnson, B.B., and Wells, J.D., 1998. The influence of temperature on the adsorption of cadmium(II) and cobalt(II) on kaolinite. *J. Colloid Interface Sci.* 204, 93-103.

Axe, L. and Anderson, P.R., 1997. Experimental and theoretical diffusivities of Cd and Sr in hydrous ferric oxide. *J. Colloid Interface Sci.* 185, 436-448.

Baeyens, B. and Bradbury, M.H., 1997. A mechanistic description of Ni and Zn sorption on Na-montmorillonite. Part I: Titration and sorption measurements. *J. Contam. Hydrol.* 27, 199-222.

Barrow, N.J., Gerth, J., and Brümmer, G.W., 1989. Reaction kinetics of the adsorption and desorption of nickel, zinc, and cadmium by goethite. II. Modelling the extent and rate of reaction. *J. Soil Sci.* 40, 437-450.

Benedetti, M.F., Milne, C.J., Kinniburgh, D.G., van Riemsdijk, W.H., and Koopal, L.K., 1995. Metal ion binding to humic substances: application of the non-ideal competitive adsorption model. *Environ. Sci. Technol.* 29, 446-457.

Benedetti, M.F., van Riemsdijk, W.H., Koopal, L.K., Kinniburgh, D.G., Gooddy, D.C., and Milne, C.J., 1996. Metal ion binding by natural organic matter: from the model to the field. *Geochim. Cosmochim. Acta.* 60, 2503-2513.

Benjamin, M.E. and Leckie, J.O., 1981a. Competitive adsorption of Cd, Cu, Zn, and Pb on amorphous iron oxyhydroxide. *J. Colloid Interface Sci.* 83, 410-419.

Benjamin, M.M. and Leckie, J.O., 1981b. Multiple-site adsorption of Cd, Cu, Zn, and Pb on amorphous iron oxyhydroxide. *J. Colloid Interface Sci.* 79, 209-221.

Bibak, A., 1997. Competitive sorption of copper, nickel, and zinc by an oxisol. *Comm. Soil Sci. Plant Anal.* 28, 927-937.

Boekhold, A.E. and van der Zee, S.E.A.T.M., 1992. A scaled sorption model validated at the column scale to predict cadmium contents in a spatially variable field soil. *Soil Sci.* 154, 105-112.

Bowden, J.W., Bolland, M., Posner, A.M., and Quirk, J.P., 1973. Generalized model for anion and cation adsorption at oxide surfaces. *Nature Phys. Sci.* 245, 81-83.

Bowden, J.W., Posner, A.M., and Quirk, J.P., 1977. Ionic adsorption on variable charge mineral surfaces. Theoretical charge development and titration curves. *Austr. J. Soil Res.* 18, 49-60.

Bradbury, M.H. and Baeyens, B., 1997. A mechanistic description of Ni and Zn sorption on Na-montmorillonite. Part II: Modelling. *J. Contam. Hydrol.* 27, 223-248.

Brüemmer, G.W., Gerth, J., and Tiller, K.G., 1988. Reaction kinetics of the adsorption and desorption of nickel, zinc and cadmium by goethite. I. Adsorption and diffusion of metals. *J. Soil Sci.* 39, 37-52.

Buchter, B., Davidoff, B., Amacher, M.C., Hinz, C., Iskander, I.K., and Selim, H.M., 1989. Correlation of Freundlich K_d and n retention parameters with soils and elements. *Soil Sci.* 148, 370-379.

Cavallaro, N. and McBride, M.B., 1978. Copper and cadmium adsorption characteristics of selected acid and calcareous soils. *Soil Sci. Soc. Am. J.* 42, 550-556.

Christensen, J.B., Tipping, E., Kinniburgh, D.G., Gron, C., and Christensen, T.H., 1998. Proton binding by groundwater fulvic acids of different age, origins, and structure modeled with the model V and NICA-Donnan model. *Environ. Sci. Technol.* 32, 3346-3355.

Christensen, T.H., 1987a. Cadmium soil sorption at low concentrations: V. Evidence of competition by other heavy metals. *Water Air Soil Poll.* 34, 293-303.

Christensen, T.H., 1987b. Cadmium soil sorption at low concentrations: VI. A model for zinc competition. *Water Air Soil Poll.* 34, 305-314.

Christensen, T.H., Lehmann, N., Jackson, T., and Holm, P.E., 1996. Cadmium and nickel distribution coefficients for sandy aquifer materials. *J. Contam. Hydrol.* 24, 75-84.

Christl, I. and Kretzschmar, R., 1999. Competitive sorption of copper and lead at the oxide-water interface: implications for surface site density. *Geochim. Cosmochim. Acta.* 63, 2929-2938.

Cowan, C.E., Zachara, J.M., and Resch, C.T., 1991. Cadmium adsorption on iron oxides in the presence of alkaline-earth elements. *Environ. Sci. Technol.* 25, 437-446.

Curtin, D. and Rostad, H.P.W., 1997. Cation exchange and buffer potential of Saskatchewan soils estimated from texture, organic matter and pH. *Can. J. Soil Sci.* 77, 621-626.

Davis, J.A., Coston, J.A., Kent, D.B., and Fuller, C.C., 1998. Application of the surface complexation concept to complex mineral assemblages. *Environ. Sci. Technol.* 32, 2820-2828.

Davis, J.A., James, R.O., and Leckie, J.O., 1978. Surface ionization and complexation at the oxide/water interface: 1. Computation of electrical double layer properties in simple electrolytes. *J. Colloid Interface Sci.* 63, 480-499.

Davis, J.A. and Leckie, J.O., 1978. Surface ionization and complexation at the oxide/water interface: 2. Surface properties of amorphous iron oxyhydroxides and adsorption of metal ions. *J. Colloid Interface Sci.* 67, 90-107.

Doner, H.E., 1978. Chloride as a factor in mobilities of Ni(II), Cu(II), and Cd(II) in soil. *Soil Sci. Soc. Am. J.* 42, 882-885.

Dzombak, D.A. and Morel, F.M.M., 1990. *Surface Complexation Modeling, Hydrous Ferric Oxide.* John Wiley & Sons, New York.

Elzinga, E.J. and Sparks, D.L., 1999. Nickel sorption mechanisms in a pyrophyllite-montmo-rillonite mixture. *J. Colloid Interface Sci.* 213, 506-512.

Elzinga, E.J., van Grinsven, J.J.M., and Swartjes, F.A., 1999. General purpose Freundlich isotherms for cadmium, copper, and zinc in soils. *Eur. J. Soil Sci.* 50, 139-149.

Ford, R.G., Scheinost, A.C., Scheckel, K.G., and Sparks, D.L., 1999. The link between clay mineral weathering and the stabilization of Ni surface precipitates. *Environ. Sci. Technol.* 33, 3140-3144.

Gaines, G.L.J. and Thomas, H.C., 1953. Adsorption studies on clay minerals. II: A formulation of the thermodynamics of exchange adsorption. *J. Chem. Phys.* 21, 714-718.

Gerth, J., Brümmer, G.W., and Tiller, K.G., 1993. Retention of Ni, Zn and Cd by Si-associated goethite. *Zeitschrift für Pflanzenernährung und Bodenkunde.* 156, 123-129.

Goldberg, S., 1992. Use of surface complexation models in soil chemical systems. *Adv. Agron.* 47, 233-329.

Goldberg, S., 1999. Reanalysis of boron adsorption on soils and soil minerals using the constant capacitance model. *Soil Sci. Soc. Am. J.* 63, 823-829.

Grolimund, D., Borkovec, M., Federer, P., and Sticher, H., 1995. Measurement of sorption isotherms with flow-through reactors. *Environ. Sci. Technol.* 29, 2317-2321.

Gutierrez, M. and Fuentes, H.R., 1991. Competitive adsorption of cesium, cobalt, and stron-tium in conditioned clayey soil suspensions. *J. Environ. Radioactivity.* 13, 271-282.

Harter, R.D., 1992. Competitive sorption of cobalt, copper, and nickel ions by a calcium-saturated soil. *Soil Sci. Soc. Am. J.* 56, 444-449.

Hayes, K.F. and Leckie, J.O., 1986. Mechanism of lead ion adsorption at the goethite-water interface. *ACS Symposium Series* 323, 114-141.

Hayes, K.F. and Leckie, J.O., 1987. Modeling ionic strength effects on cation adsorption at hydrous oxide/solution interfaces. *J. Colloid Interface Sci.* 115, 564-572.

Hendershot, W.H. and Duquette, M., 1986. A simple barium chloride method for determining cation exchange capacity and exchangeable cations. *Soil Sci. Soc. Am. J.* 50, 605-608.

Hering, J.G. and Morel, F.M.M., 1988. Humic acid complexation of calcium and copper. *Environ. Sci. Technol.* 22, 1234-1237.

Hesterberg, D., Bril, J., and del Castilho, P., 1993. Thermodynamic modeling of zinc, cadmium, and copper solubilities in a manured, acidic, loamy-sand topsoil. *J. Environ. Qual.* 22, 681-688.

Hiemstra, T., de Wit, J.C.M., and van Riemsdijk, W.H., 1989a. Multisite proton adsorption modeling at the solid/solution interface of (hydr)oxides: a new approach. II. Application to various important (hydr)oxides. *J. Colloid Interface Sci.* 133, 105-117.

Hiemstra, T. and van Riemsdijk, W.H., 1996. A surface structural approach to ion adsorption: the charge distribution (CD) model. *J. Colloid Interface Sci.* 179, 488-508.

Hiemstra, T., van Riemsdijk, W.H., and Bolt, G.H., 1989b. Multisite proton adsorption modeling at the solid/solution interface of (hydr)oxides: a new approach. I. Model description and evaluation of intrinsic reaction constants. *J. Colloid Interface Sci.* 133, 91-104.

Hoins, U., Charlet, L., and Sticher, H., 1993. Ligand effect on the adsorption of heavy metals: the sulfate-cadmium goethite case. *Water Air Soil Poll.* 68, 241-255.

Huang, C.P. and Stumm, W., 1973. Specific adsorption of cations on hydrous γ-Al_2O_3. *J. Colloid Interface Sci.* 43, 409-420.

Jenne, E.A., 1998. Adsorption of metals by geomedia: data analysis, modeling, controlling factors, and related issues. In *Adsorption of Metals by Geomedia.* Jenne, E.A., Ed., Academic Press, San Diego.

Karthikeyan, K.G., Elliot, H.A., and Chorover, J., 1999. Role of surface precipitation in copper sorption by the hydrous oxides of iron and aluminum. *J. Colloid Interface Sci.* 209, 72-78.

Keizer, M.G., De Wit, J.C., Meussen, J.C.L., Bosma, W.J.P., Nederlof, M.M., Venema, P., Meussen, V.C.S., van Riemsdijk, W.H., and van der Zee, S.E.A.T.M., 1993. ECOSAT: a computer program for the calculation of speciation in soil-water systems. Wageningen, The Netherlands.

Kinniburgh, D.G., Benedetti, M.F., van Riemsdijk, W.H., and Kretzschmar, R. (2000). Fundamental aspects of metal speciation and transport in metal-contaminated soils and aquifers (FAMEST): Second Annual Report, Rep. No. WD/00/02. British Geological Survey, Keyworth, Nottinghamshire.

Kinniburgh, D.G., Milne, C.J., Benedetti, M.F., Pinheiro, J.P., Filius, J., Koopal, L.K., and van Riemsdijk, W.H., 1996. Metal ion binding by humic acid: application of the NICA-Donnan model. *Environ. Sci. Technol.* 30, 1687-1698.

Kinniburgh, D.G., van Riemsdijk, W.H., Koopal, L.K., and Benedetti, M.F., 1998. Ion binding to humic substances: measurements, models, and mechanisms. In *Adsorption of Metals by Geomedia.* Jenne, E.A., Ed., Academic Press, San Diego.

Kinniburgh, D.G., van Riemsdijk, W.H., Koopal, L.K., Borkovec, M., Benedetti, M.F., and Avena, M.J., 1999a. Ion binding to natural organic matter: competition, heterogeneity, stoichiometry and thermodynamic consistency. *Colloids and Surfaces A — Physicochemical and Engineering Aspects.* 151, 147-166.

Kraepiel, A.M., Keller, K., and Morel, F.M.M., 1998. On the acid-base chemistry of permanently charged minerals. *Environ. Sci. Technol.* 32, 2829-2838.

Kraepiel, A.M.L., Keller, K., and Morel, F.M.M., 1999. A model for metal adsorption on montmorillonite. *J. Colloid Interface Sci.* 210, 43-54.

Lee, S.-Z., Allen, H.E., Huang, C.P., Sparks, D.L., Sanders, P.F., and Peijnenburg, W.J.G.M., 1996. Predicting soil-water partition coefficients for cadmium. *Environ. Sci. Technol.* 30, 3418-3424.

Lützenkirchen, J., 1997. Ionic strength effects on cation sorption to oxides: macroscopic observations and their significance in microscopic interpretation. *J. Colloid Interface Sci.* 195, 149-155.

Mandal, R., Salam, M.S.A., Murimboh, J., Hassan, N.M., Chakrabarti, C.L., Back, M.H., and Gregoire, D.C., 2000. Competition of Ca(II) and Mg(II) with Ni(II) for binding by a well-characterized fulvic acid in model solutions. *Environ. Sci. Technol.* 34, 2201-2208.

McBride, M.B., 1989. Reactions controlling heavy metal solubility in soils. *Adv. Soil Sci.* 10, 1-56.

McKnight, D.M. and Wershaw, R.L., 1994. In *Humic Substances in the Suwannee River: Interactions, Properties, and Proposed Structures*. Averett, R.C., Leenheer, J.A., McKnight, D.M.,Thorn, K.A., Eds., USGS Water-Supply Paper 2373, United States Geological Survey, Denver.

Milne, C.J., Kinniburgh, D.G., de Wit, J.C.M., van Riemsdijk, W.H., and Koopal, L.K., 1995. Analysis of metal-ion binding by a peat humic acid using a simple electrostatic model. *J. Colloid Interface Sci.* 175, 448-460.

Morin, G., Ostergren, J.D., Juillot, F., Ildefonse, P., Calas, G., and Brown, G.E.J., 1999. XAFS determination of the chemical form of lead in smelter-contaminated soils and mine tailings: importance of adsorption processes. *Am. Mineralogist.* 84, 420-434.

Palmqvist, U., Ahlberg, E., Lövgren, L., and Sjöberg, S., 1999. Competitive metal ion adsorption in goethite systems using *in situ* voltammetric methods and potentiometry. *J. Colloid Interface Sci.* 218, 388-396.

Papini, A.P., Kahie, Y.D., Troia, B., and Majone, M., 1999. Adsorption of lead at variable pH onto a natural porous medium: modeling of batch and column experiments. *Environ. Sci. Technol.* 33, 4457-4464.

Paulson, A.J. and Balistrieri, L., 1999. Modeling removal of Cd, Cu, Pb, and Zn in acidic groundwater during neutralization by ambient surface waters and groundwaters. *Environ. Sci. Technol.* 33, 3850-3856.

Pinheiro, J.P., Mota, A.M., and Benedetti, M.F., 1999. Lead and calcium binding to fulvic acids: salt effect and competition. *Environ. Sci. Technol.* 33, 3398-3404.

Puls, R.W., Powell, R.M., Clark, D., and Eldred, C.J., 1991. Effects of pH, solid/solution ratio, ionic strength, and organic acids on Pb and Cd sorption on kaolinite. *Water Air Soil Poll.* 57-58, 423-430.

Roberts, D.R., Scheidegger, A.M., and Sparks, D.L., 1999. Kinetics of mixed Ni-Al precipitate formation on a soil clay fraction. *Environ. Sci. Technol.* 33, 3749-3754.

Robertson, A.P. and Leckie, J.O., 1998. Acid/base, copper binding, and Cu^{2+}/H^+ exchange properties of goethite, an experimental and modeling study. *Environ. Sci. Technol.* 32, 2519-2530.

Rustad, J.R., Felmy, A.R., and Hay, B.P., 1996. Molecular statics calculations of proton binding to goethite surfaces: a new approach to estimation of stability constants for multisite surface complexation models. *Geochim. Cosmochim. Acta.* 60, 1563-1576.

Sauve, S., Hendershot, W., and Allen, H.E., 2000a. Solid-solution partitioning of metals in contaminated soils: dependence on pH, total metal burden, and organic matter. *Environ. Sci. Technol.* 34, 1125-1131.

Sauve, S., Norvell, W.A., McBride, M., and Hendershot, W., 2000b. Speciation and complexation of cadmium in extracted soil solutions. *Environ. Sci. Technol.* 34, 291-296.

Scheidegger, A.M., Fendorf, M., and Sparks, D.L., 1996. Mechanisms of nickel sorption on pyrophyllite: macroscopic and microscopic approaches. *Soil Sci. Soc. Am. J.* 60, 1763-1772.

Scheidegger, A.M., Strawn, D.G., Lamble, G.M., and Sparks, D.L., 1998. The kinetics of mixed Ni-Al hydroxide formation on clay and aluminium oxide minerals: a time-resolved XAFS study. *Geochim. Cosmochim. Acta.* 62, 2233-2245.

Schindler, P.W. and Gamsjäger, H., 1972. Acid-base reactions of TiO2 (anatase)-water interface and the point of zero charge of TiO_2 suspensions. *Kolloid-Z. Z. Polym.* 250, 759-763.

Schindler, P.W. and Kamber, H.R., 1968. Die Acidität von Silanolgruppen. *Helv. Chim. Acta.* 51, 1781-1786.

Schindler, P.W., Liechti, P., and Westall, J.C., 1987. Adsorption of copper, cadmium and lead from aqueous solution to the kaolinite/water interface. *Netherlands J. Agric. Sci.* 35, 219-230.

Sparks, D.L., 1988. *Kinetics of Soil Chemical Processes.* Academic Press, San Diego.

Sparks, D.L., 1995. *Environmental Soil Chemistry.* Academic Press, San Diego.

Strawn, D.G. and Sparks, D.L., 2000. Effects of soil organic matter on the kinetics and mechanisms of Pb(II) sorption and desorption in soils. *Soil Sci. Soc. Am. J.* 64, 144-156.

Stumm, W., Huang, C.P., and Jenkins, S.R., 1970. Specific chemical interaction affecting the stability of dispersed systems. *Croatica Chemica Acta.* 42, 223-244.

Sverjensky, D.A. and Sahai, N., 1996. Theoretical prediction of single-site surface-protonation equilibrium constants for oxides and silicates in water. *Geochim. Cosmochim. Acta.* 60, 3773-3797.

Temminghoff, E.J.M., van der Zee, S.E.A.T.M., and de Haan, F.A.M., 1995. Speciation and calcium competition effects on cadmium sorption by sandy soils at various pH levels. *Eur. J. Soil Sci.* 46, 649-655.

Temminghoff, E.J.M., van der Zee, S.E.A.T.M., and de Haan, F.A.M., 1997. Copper mobility in a copper-contaminated sandy soil as affected by pH and solid and dissolved organic matter. *Environ. Sci. Technol.* 31, 1109-1115.

Thompson, H.A., Parks, G.A., and Brown, G.E., Jr., 2000. Formation and release of cobalt(II) sorption and precipitation products in aging kaolinite-water slurries. *J. Colloid Interface Sci.* 222, 241-253.

Tipping, E., 1993. Modeling the competition between alkaline earth cations and trace metal species for binding by humic substances. *Environ. Sci. Technol.* 27, 520-529.

Tipping, E., 1998. Humic ion binding model VI: an improved description of the interactions of protons and metal ions with humic substances. *Aquat. Geochem.* 4, 3-48.

Tipping, E., Fitch, A., and Stevenson, F.J., 1995. Proton and copper binding by humic acid: application of a discrete-site/electrostatic ion-binding model. *Eur. J. Soil Sci.* 46, 95-101.

Tipping, E. and Hurley, M.A., 1992. A unifying model of cation binding by humic substances. *Geochim. Cosmochim. Acta.* 56, 3627-3641.

Town, R.M. and Powell, H.K.J., 1993. Ion-selective electrode potentiometric studies on the complexation of copper(II) by soil-derived humic and fulvic acids. *Analytica Chimica Acta.* 279, 221-233.

van Riemsdijk, W.H., Bolt, G.H., Koopal, L.K., and Blaakmeer, J., 1986. Electrolyte adsorption on heterogeneous surfaces: adsorption models. *J. Colloid Interface Sci.* 109, 219-228.

Venema, P., Hiemstra, T., and van Riemsdijk, W.H., 1996. Comparison of different site binding models for cation sorption: description of pH dependency, salt dependency, and cation-proton exchange. *J. Colloid Interface Sci.* 181, 45-59.

Voegelin, A. and Kretzschmar, R., 2001. Competitive sorption and transport of Cd(II) and Ni(II) in soil columns: application of a cation exchange equilibrium model (submitted).

Voegelin, A., Vulava, V.M., and Kretzschmar, R., 2001. Reaction-based model describing competitive sorption and transport of Cd, Zn, and Ni in an acidic soil. *Env. Sci. Technol.* 35, 1651-1657.

Voegelin, A., Vulava, V.M., Kuhnen, F., and Kretzschmar, R., 2000. Multicomponent transport of major cations predicted from binary sorption experiments. *J. Contam. Hydrol.* 46, 319-338.

Wada, K. and Abd-Elfattah, A., 1979. Effects of cation exchange material on zinc adsorption by soils. *J. Soil Sci.* 30, 281-290.

Wang, F., Chen, J., and Forsling, W., 1997a. Modeling sorption of trace metals on natural sediments by surface complexation model. *Environ. Sci. Technol.* 31, 448-453.

Wang, W.-Z., Brusseau, M.L., and Artiola, J.F., 1997b. The use of calcium to facilitate desorption and removal of cadmium and nickel in subsurface soils. *J. Contam. Hydrol.* 25, 325-336.

Wang, W.-Z., Brusseau, M.L., and Artiola, J.F., 1998. Nonequilibrium and nonlinear sorption during transport of cadmium, nickel, and strontium through subsurface soils. In *Adsorption of Metals by Geomedia*. Jenne, E.A., Ed., Academic Press, San Diego.

Wen, X., Du, Q., and Tang, H., 1998. Surface complexation model for the heavy metal adsorption on natural sediment. *Environ. Sci. Technol.* 32, 870-875.

Westall, J.C., 1986. Reactions at the oxide-solution interface: Chemical and electrostatic models. In *Geochemical Processes at Mineral Surfaces*. American Chemical Society, Washington, D.C.

Xia, K., Mehadi, A., Taylor, R.W., and Bleam, W.F., 1997. X-ray absorption and electron paramagnetic resonance studies of Cu(II) sorbed to silica: surface-induced precipitation at low surface coverages. *J. Colloid Interface Sci.* 185, 252-257.

Yates, D.E., Levine, S., and Healy, T.W., 1974. Site-binding model of the electrical double layer at the oxide/water interface. *J. Chem. Soc. Far. Trans.* 70, 1807-1818.

Zachara, J.M., Smith, S.C., Resch, C.T., and Cowan, C.E., 1992. Cadmium sorption to soil separates containing layer silicates and iron and aluminum oxides. *Soil Sci. Soc. Am. J.* 56, 1074-1084.

Zasoski, R.J. and Burau, R.G., 1988. Sorption and sorptive interaction of cadmium and zinc on hydrous manganese oxide. *Soil Sci. Soc. Am. J.* 52, 81-87.

CHAPTER 4

Heavy Metal Solubility and Transport in Soil Contaminated by Mining and Smelting

Steven L. McGowen and Nicholas T. Basta

INTRODUCTION

Metal mining, smelting, and processing throughout the world have contaminated soils with heavy metals in excess of natural soil background concentrations. These processes introduce metal contaminants into the environment through gaseous and particulate emissions, waste liquids, and solid wastes (Dudka and Adriano, 1997). In addition to the soil contamination from these pathways, many mining and smelting sites have considerable surface water and groundwater contamination from heavy metals released and transported from contaminated soils. This contamination endangers water supply resources as well as the economic and environmental health of surrounding communities.

Many metal solubility and transport studies in soil have centered on land application of wastes, such as biosolids, that contain metal contaminants. Dowdy and Volk (1983) presented an excellent review of metal movement in soils amended with sewage sludge. Principles behind the chemistry, solubility, and potential movement of metals in soils have been presented in excellent reviews by McLean and Bledsoe (1992) and McBride (1989). In addition to soil chemical reactions that affect metal solubility and potential movement, other studies have provided information on modeling metal transport through soil systems. The use of models to describe transport and retention of metals through soils and the relationship between transport and soil physical and chemical properties has been reviewed (Selim, 1992). Research techniques and methods for quantifying metal retention and transport in uncontaminated soils have been reviewed (Selim and Amacher, 1997).

Research involving heavy metal retention and transport and the quantification of these parameters has largely been undertaken using uncontaminated soils and the addition of metal salt reagents to simulate chemical and physical systems of contaminated sites. While not an accurate representation of "real world" sites contaminated with mine and smelting wastes, results from such experiments have provided technical expertise and an abundance of detailed information on the chemical, physical, and transport processes involved with metal–soil interactions. With this wealth of knowledge, many researchers are now designing research methods to apply related techniques to soils contaminated from mining, smelting, and refining of metal ores.

In this chapter, metal release and transport from soils contaminated from anthropogenic activities related to metal mining, extraction, and processing will be reviewed. The first part of this review focuses on research approaches designed to study the solubility and transport of metals from soils contaminated by metal exploration, processing, and smelting of ores. The second section of this chapter focuses on novel techniques and methods designed to evaluate the efficiency of chemical treatments for reducing metal release and transport from contaminated soils.

METHODS USED TO QUANTIFY RELEASE AND TRANSPORT OF METALS IN CONTAMINATED SOILS

The methods used to investigate the solubility and mobility of metals from mine- and smelter-contaminated soils include measuring the distribution of metal contaminants with progressive depth in soil profiles, monitoring the mineralogical properties of contaminants to determine potential for dissolution and transport, the use of chemical partitioning methods to indicate soil chemical constituents responsible for sorption and release of metals, and the use of leaching experiments and water monitoring in the environment to determine the impact of specific contamination regimes. Each type of investigative method has the common objective of describing or interpreting the means of metal release, mobility, and transport from contaminated soils. Most research studies have focused on one or more of the following approaches: (1) soil profile metal distribution, (2) mineralogical properties, (3) chemical partitioning, and (4) leaching and water monitoring (Table 4.1).

Soil Profile Metal Distribution and Metal Transport

The distribution of metal contaminants within the depth profile of soil can indicate the relative mobility of metals originating from surface deposited contamination. Soil profile metal distribution studies often involve deep soil core sampling with incremental division of the cores at specific depths for chemical analyses. The depth profile distribution of metals and other soil chemical properties is then investigated to determine soil chemical properties that contribute to metal retention or transport.

Scocart et al. (1983) presented soil profile distributions of Cd, Pb, Zn, and Cu for two soil types (sandy acidic soils and loamy neutral soils) that were taken directly adjacent to or at 2 km upwind from a Zn smelter. Total metal content was determined from HCl-HNO_3 digestion, and exchangeable metal was determined by $1 N NH_4$-acetate extractions at a 1:10 soil:solution ratio. For both soil types, total metal concentration was greater in soils sampled closer to the smelter site. For the sandy soils,

Table 4.1 Author, Metals Investigated, Contaminant Source, and Method of Investigation of Selected Articles

Author	Metals	Contaminant Source	Method of Investigation
Scocart et al., 1983	Cd, Cu, Pb, Zn	Smelter emissions	Soil profile metal distribution
Wilkens and Loch, 1997	Cd, Zn	Smelter emissions	Soil profile metal distribution
Maskall, et al., 1994	Cd, Pb, Zn	Smelter waste	Soil profile metal distribution
Cernik et al., 1994	Cu, Zn	Smelter waste	Soil profile metal distribution/modeling
Gee et al., 1997	Pb	Pb smelter waste	Mineralogical
Mattigod et al., 1986	Cd, Pb, Zn	Mine waste	Mineralogical
Hagni and Hagni, 1991	Cu, Pb, Zn	Smelting waste	Mineralogical
Li and Shuman, 1996	Cd, Pb, Zn	Refining waste	Chemical partitioning/soil profile metal distribution
Abdel-Saheb et al., 1994	Cd, Pb, Zn	Mine tailings	Chemical partitioning
Fanfani et al., 1997	Pb, Zn	Mine tailings	Chemical partitioning/mineralogical
Benner et al., 1995	As, Cd, Cu, Fe, Pb, Zn	Mine tailings	Leaching and water monitoring/mineralogical
Fuge et al. 1993	Cd, Zn	Mine tailings	Water monitoring
Paulson, 1997	Cd, Cu, Fe, Mn, Pb, Zn	Mine tailings	Water monitoring
Shackleford and Glade, 1997	Cd, Pb, Zn	Refining waste	Leaching/modeling
Zhu et al., 1999	Cd, Pb, Zn	Mine tailings	Unsaturated leaching

the distribution of total Cd and Zn within the soil profiles was correlated to the organic matter of the soil, suggesting that metal–organic complexes are responsible for their movement to lower depths (20 to 30 cm). Furthermore, distribution of Cu and Pb were mostly concentrated in the upper surface layers (0 to 10 cm). For the loamy soils at neutral pH, the majority of the heavy metals remained in the upper layers of the soil, but at decreased pH (5.7), dissolution of metals and movement to lower depths in the soil profiles was observed.

Wilkens and Loch (1997) also investigated the distribution of Cd and Zn from smelter emissions in acid soils. Their research centered on soil chemical components potentially responsible for retention and/or movement of metals in soils. These included aluminum, iron, and manganese oxides, organic matter, clay, carbonates, and soil pH. By sampling the soil profile at incremental depths, they determined that the organic fraction was the fraction best correlated with Cd and Zn distribution. The effect of low pH in the study soils was also noted to have decreased the retention of metals by the oxide fractions due to pH values below the point of zero charge of these oxide fractions. In another study of metal migration into soils, Maskall et al. (1994) sampled soil cores from five historic smelting sites. Cores were taken in sites with heavy surface contamination of Pb, Cd, and Zn from slag wastes. Their work concluded that metal mobility was slowed in clay soils by CEC with Pb migration rates of (0.07 to 0.32 cm yr^{-1}). However, migration rates were much higher (0.78 cm yr^{-1}) at sites dominated by sandstone geology due to preferential flow along cracks and fissures in the parent rock.

Cernik et al. (1994) observed soil depth profiles of Zn and Cu originating from smelter airborne pollution. Along with the observed metal distribution in soil profiles, the history of metal production at the site was taken into consideration for quantitatively describing metal distribution with transport models. By using linear adsorption isotherms generated in laboratory experiments for parameter estimation, they were able to generally describe Zn depth profiles using the convection-dispersion equation with no fitting parameters. When using the linear convection-dispersion model, the agreement of the model with the experimental data was significantly improved by varying chosen values of dispersivity. In addition to fitting observed depth profiles of Zn and Cu, Cernik et al. (1994) also calculated predicted future depth profiles using transport models and their change as effected by hypothetical remediation techniques. Their predictions were based on four treatment strategies: (1) no action, (2) plow to 20 cm, (3) removal of the top 5 cm, or (4) removal of the top 10 cm and replacement with uncontaminated soil. Based on these treatments, removal and replacement of the top 10 cm of soil resulted in the most improved predicted metal depth profile.

With diverse site-specific conditions (such as source of contaminant, co-contaminant chemistry, precipitation, soil type and chemistry, geology, etc.), each description of metal transport using the technique of soil profile metal distribution is inherently unique. This method of describing metal transport may therefore limit the broad interpretation of results to other sites with dissimilar characteristics. Even so, detailed site-by-site description and determination of transport-related processes responsible for redistributing metal contaminants gives valuable information for remediation efforts at high-priority locations.

Mineralogical Properties and Metal Transport

Mineralogical methods can be used to estimate potential release and transport of metals from mining and smelting wastes. In this approach, potential solubility and mobility of a metal contaminant is estimated from the mineral form of the heavy metal contaminant. When specific mineral forms are determined, the solubility under varying soil and environmental conditions can be implied and related to potential release and transport. The underlying assumption is that the metal concentration is controlled by the solubility product (K_{SP}) of the mineral. Solubility products are used to compute the concentration of a solute in equilibrium with a solid (precipitate or mineral) phase. By applying the solubility product concept to waste products of mineral composition in smelter or mining wastes, wastes with lower K_{SP} values should be less soluble and therefore decrease the potential for metal release and transport.

Trace metal solid phases can be determined indirectly or directly. Indirect methods are qualitative predictions based on solution species activities and their input into chemical equilibrium models or the plotting of activities on solubility, phase, or activity-ratio diagrams. Direct methods use spectroscopic instruments such as X-ray diffraction (XRD) or scanning electron microscopy coupled with energy dispersive X-ray analysis (SEM-EDX) to quantify mineral components. To avoid the expense of the direct measuring instruments, indirect methods may be advantageous for gaining insight into potential minerals controlling solubility in heterogeneous noncrystalline materials. However, indirect methods do not provide the explicit evidence of definite mineral identification, as is the case with using direct spectroscopic methods.

Gee et al. (1997) investigated the mineralogy of smelting slags and the response of specific mineral forms to natural weathering processes and potential release into the environment. They collected several slag samples from historic smelting centers of the Roman age (100 to 200 A.D.) and the medieval age (1300 to 1550) in the U.K. and subjected them to analysis using SEM-EDX. Several lead phases such as lead oxide, pyromorphite, cerrusite, hydrocerrusite, galena, leadhillite, and anglesite were positively identified by XRD. Of these crystalline Pb phases, the presence of cerrusite, hydrocerrusite, and pyromorphite were identified as being weathered forms of primary lead minerals. The authors also noted that weathering of unstable minerals such as dicalcium silicates were buffering the slag-soil system pH and moderating the effect of naturally acidic rainwater to dissolve heavy metal minerals. This buffering effect was also attributed to slowing downward migration of Pb into groundwater.

Other researchers have used spectroscopic techniques to investigate and identify metal mineral phases in contaminated soils. Mattigod et al. (1986) identified mineral fractions from a mine-waste contaminated soil using size and density fractionation coupled with XRD and SEM-EDX analyses. Hagni and Hagni (1991) identified the mineralogy of wastes from lead smelting wastes using reflected light microscopy and SEM-EDX. These and many other studies have broadened the understanding between the relationship of metal mineral phase and metal solubility as related to transport.

In addition to the mineralogy present on contaminated sites, individual site factors are also important. The effect of oxidation state on the solubility of metal sulfate minerals is well known (Nordstrom, 1982). The production of acidic and metal-enriched waters from the oxidation of sulfide ores and wastes often occurs around smelting and mining sites. Consideration of mineralogical control on metal solubility and transport must take into account the effects of redox, weathering environment, co-contaminants, and site conditions. Furthermore, mineralogical methods cannot provide quantitative analyses of mine- and smelter-contaminated soils for all situations. Many mine and smelter sites contain amorphous waste materials that are very difficult or impossible to characterize using direct mineralogical methods. Release and potential transport of metals from amorphous wastes must be investigated using indirect mineralogical methods or other techniques.

Chemical Partitioning and Metal Transport

The use of chemical partitioning or sequential fractionation to determine the concentration of metals present in specific soil chemical fractions has been used to determine plant availability (Chlopecka and Adriano, 1996), plant and human availability (Basta and Gradwohl, 2000), and heavy metal movement in soil profiles (Li and Shuman, 1996). In many sequential fractionation schemes, the strength of salts and acids increase incrementally for successive extractions of the same soil sample. Often the fraction referred to as "potentially mobile" is the solution obtained from a weak salt or deionized water extraction. Other fractions are often operationally defined as organic, oxide-bound, or residual fractions. These fractions are typically considered immobile unless specific environmental conditions are induced in the soil environment.

Chemical partitioning of metal fractions to determine metal release and transport from mine- and smelter-waste has been studied by many researchers. Li and Shuman (1996) used a sequential extraction procedure to investigate the movement of metal fractions from a steel production waste into the soil profile. Their scheme divided the sequential extractions into five phases: exchangeable, organic, Mn-oxide, amorphous Fe-oxide, crystalline Fe-oxide, and residual. Soils were sampled at incremental depths within the profile and subjected to the sequential fractionation scheme. For the distribution of metal in the soil profile below 30 cm, the exchangeable Zn fraction became the dominant fraction. Furthermore, soil samples at 100 cm indicated exchangeable Zn, which indicated that Zn may enter the shallow water table of some of the soils in the study. Compared with Zn, Cd and Pb movement in the soil profiles was minimal. This minimal movement was attributed to Cd and Pb sorption to the organic fraction, which prevented excessive downward movement into the soil profile.

Abdel-Saheb et al. (1994) also used a sequential extraction scheme to understand the plant availability and runoff loss potential of heavy metals mine tailings. Their research attributed the exchangeable (0.01 M CaCl$_2$) and sorbed (water soluble) fractions as being the most susceptible to plant uptake and leaching. Their results indicated that Zn was relatively immobile near the tailing piles but increased in mobility with distance away from the tailing piles. This phenomenon was attributed to the high pH present in the tailing piles that buffered the system and decreased Zn solubility. Solubility of Zn increased away from the piles due to a decrease in soil pH with distance from the tailing piles. Greatest concentrations of Cd, Pb, and Zn were found in the sulfide fraction (4 M HNO_3 extraction), indicating that a large percentage of these metals were present as sulfide minerals with little weathering or oxidation since extraction as ore.

Fanfani et al. (1997) also used speciation analysis and spectroscopic methods (XRD and SEM-EDX) to investigate Pb and Zn mine tailings weathering and transport of metals. Their fractionation scheme indicated that Zn and Cd were easily dissolved from the waste and that Pb was relatively insoluble. These results corresponded with observed stream water analyses near the vicinity of the waste impoundments. Fanfani et al. (1997) also cited two important advantages of investigating weathering processes through chemical partitioning: (1) the analyses did not require mineralogical laboratory equipment, and (2) the partitioning scheme allowed for investigation of trace elements that are amorphous in composition and difficult to investigate using mineralogical methods but still have potential environmental impact. These two factors essentially summarize the advantages using chemical partitioning to investigate metal release and transport in contaminated soils. Often where direct mineralogical methods fail, such as with amorphous waste materials, chemical partitioning methods can provide a wealth of information on the potential release and transport of metal contaminants.

Column Transport Studies, Groundwater Monitoring, and Metal Transport

Highly contaminated land exposed to natural weathering processes has dispersed metal contaminants beyond historic boundaries to surrounding soils, streams, and groundwater (Fuge et al., 1993; Paulson, 1997). The redistribution of metal contaminants through leaching and surface transport processes endangers the quality of waters used for human consumption and threatens the welfare of surrounding ecosystems.

To investigate the release of metals from mine tailings and transport to water resources, Benner et al. (1995) used aluminum silicate ceramic beads installed *in situ* to collect mineral coatings in addition to groundwater sampling. Their research showed that the hyporheic zone (zone of mixed ground and surface water) may act as a sink for metals. Furthermore, results showed that metal loading to the stream bank sediment from mine tailings weathering was highly significant.

Extensive work has been done to model metal transport through uncontaminated soils (Jurinak and Santillian-Medrano, 1974; Selim et al., 1990; Selim, 1992; Selim and Amacher, 1997). Methods used for these studies involved the addition of metal salt solutions via pulse or continuous flow through repacked uncontaminated soils until metal breakthrough or complete miscible displacement. Research on the release and transport of heavy metals from contaminated soils has not been as extensive. With the wide-ranging environmental, physical, and chemical properties associated with smelting and mining waste, metal solubility from such heterogeneous wastes is also highly variable. This type of variability in metal release has made interpretation of traditional column leaching studies difficult to interpret when metal-contaminated soils are used. To address this problem, Shackelford and Glade (1997) introduced a model based on the cumulative mass of metal in the leachate instead of instantaneous metal concentration commonly used in traditional transport models. Because their model was based on the cumulative mass of metal leached, it allows for collection of large solution fractions that simplify the experimental method and reduce the number of samples for chemical analyses. Although not extensively tested, their model did work well for describing Cd, Pb, and Zn release and transport from a fly-ash amended soil.

Zhu et al. (1999) used column leaching studies to investigate the impact of soil cover and plants on metal transport in a mine-tailings contaminated soil. Their work used 15-cm-diameter columns of 60, 90, or 120 cm length to investigate combinations of clean topsoil, mine tailings, subsoil, and plant species on the release and transport of metals. Columns were leached using unsaturated conditions for 1 year. The presence of grass vegetation on the columns increased the Cd and Zn concentration in the leachate due to chemical alteration in the rhizosphere; however, the presence of subsoil acted as a sink for these metals. Lead leaching was not affected by presence of vegetation likely due to the lower solubility of Pb minerals present in the mine tailings.

The investigation of metal leachate losses and monitoring of water quality is important for defining smelting and mining sites that have high priority for remedial actions (i.e., national priority list sites). With the extensive surface and groundwater contamination present at many sites, there is a great potential for surface and subsurface transport of heavy metals with serious implications for ecosystem and human health.

METHODS FOR EVALUATING CHEMICAL TREATMENTS FOR THE REDUCTION OF HEAVY METAL MOBILITY AND TRANSPORT IN CONTAMINATED SOILS

Introduction

Cleanup of contaminated sites and disposal of metal-laden wastes are costly endeavors. Logan (1992) outlined both engineering and ecological approaches to land reclamation of chemically degraded soils. Lambert et al. (1994) also identified chemical methods to remediate metal-contaminated soils. Many remediation methods involve soil excavation, or *in situ* treatments including immobilization, mobilization, burial, washing, etc. Although highly effective at lowering risk, remediation technologies based on the excavation, transport, and landfilling of metal contaminated soils and wastes are expensive. More cost-effective techniques treat contaminants in place; however, some of these methods may temporarily exacerbate environmental risks. Soil washing increases metal solubility and mobility to remove metals from contaminated soil profiles. Increasing metal mobility for soil washing of contaminants may also increase the risk for transport and redistribution of contamination to underlying soil and groundwater (Vangronsveld and Cunningham, 1998). Other *in situ* techniques, such as vitrification, are often not feasible cleanup methods due to the high costs of energy needed to complete the process. *In situ* chemical immobilization is a remediation technique that involves the addition of chemicals to contaminated soil to form less soluble and less mobile metal compounds. Reaction products from chemical immobilization treatments are less soluble and mobile, consequently reducing heavy metal release and transport from contaminated soils to surface and groundwater. Compared with other remediation techniques, *in situ* chemical immobilization is less expensive than other remediation techniques and may provide a long-term remediation solution through the formation of low solubility metal minerals and/or precipitates.

Many of the methods used to determine the release and mobility of metals in contaminated soils can be applied to investigate the efficiency of chemical treatments to reduce heavy metal mobility and transport in mine- and smelter-contaminated soils. The following sections discuss techniques used to investigate the ability of chemical treatments to reduce metal solubility and transport.

Reduction in Metal Solubility from Inorganic Chemical Treatments

Several types of inorganic chemical treatments have been used to reduce metal solubility in soils. These treatments have been evaluated predominantly using chemical partitioning and mineralogical methods. Alkaline materials used as chemical immobilization treatments include calcium oxides, calcium and magnesium carbonates (limestone), and industrial by-products such as cement kiln dust and alkaline fly ash. Alkaline amendments can reduce heavy metal solubility in soil by increasing soil pH and metal sorption to soil particles (Filius et al., 1998; McBride et al., 1997). Increased sorption of metals to soil colloids can decrease mobile metals in solution and reduce metal transport in contaminated soils. Additionally, increased soil pH and carbonate buffering can allow the formation of metal-carbonate precipitates, complexes, and secondary minerals (Chlopecka and Adriano, 1996; McBride, 1989). Metal-carbonate minerals formed with addition of carbonate-rich limestone can decrease heavy metal solubility and reduce metal mobility and transport.

Phosphate chemical addition to contaminated soils has proven to be extremely effective for reducing metal solubility. Experiments involving treatment of metal contaminated soils with rock phosphates (apatite and hydroxyapatite) have shown that formation of metal-phosphate precipitates and minerals reduced heavy metal solubility. Insoluble and geochemically stable lead pyromorphites such as hydroxypyromorphite $[Pb_5(PO_4)_3OH]$ and chloropyromorphite $[Pb_5(PO_4)_3Cl]$ have been found to control Pb solubility in apatite amended contaminated soils (Chen et al., 1997; Eighmy et al., 1997; Laperche et al., 1997; Ma et al., 1993,1995; Ma and Rao, 1997; Zhang and Ryan, 1999).

Research of phosphate sources with higher solubility than rock phosphate (i.e., phosphate salts) has been shown to increase the efficiency of lead pyromorphite formation and reduction in metal solubility (Cooper et al., 1998; Hettiarachchi et al., 1997; Ma et al., 1993; Pierzynski and Schwab, 1993). Ma and Rao (1997) suggested that P sources with higher solubility could be mixed with rock phosphate to increase the effectiveness of lead immobilization in contaminated soils. Soluble phosphate has been shown to reduce Cd and Pb solubility (Santillian-Medrano and Jurinak, 1975). Other soluble phosphates have been shown to induce the formation of heavy metal phosphate precipitates. Materials such as Na_2HPO_4 (Cotter-Howells and

Capron, 1996) and pyrophosphate (Xie and MacKenzie, 1990) are highly effective for forming precipitates and increasing sorption Pb and Zn. Phosphorus fertilizer materials have also been tested as chemical immobilization treatments. Research with diammonium phosphate [$(NH_4)_2HPO_4$] (DAP) has shown decreased Cd solubility in soil cadmium suspensions (Levi-Minzi and Petruzzelli, 1984).

Incorporation of diammonium phosphate (DAP) reduced Cd, Pb, and Zn release and transport from a smelter-contaminated soil under saturated flow conditions in solute transport experiments (McGowen, 2000; McGowen et al., in press). Concentrations of metal species in solution fractions collected from transport experiments were input into a chemical equilibrium model to determine metal mobility as controlled by solution speciation. By assuming that negative charges dominate soil particle surfaces, low-mobility species were defined as cationic species with +1 or +2 valence (M^{2+}, MOH^+), and high-mobility species were defined as uncharged and/or anionic species with −1 or −2 valence [MSO_4^0, $M(SO_4)_2^{2-}$, etc.]. Chemical speciation of equilibrated soil solutions revealed an increase in the percentage of high-mobility metal species (anionic and uncharged species) with increased DAP treatment (Table 4.2). The increase in the percentage of high-mobility metal species with phosphate addition appears to indicate that DAP treatments may increase heavy metal mobility through soil. However, closer inspection has revealed that the concentration of high-mobility species decreased with increasing DAP treatment (Table 4.2). Therefore, treatment of contaminated soil with DAP reduced the total concentration of anionic or uncharged dissolved metal chemical species in solution and decreased heavy metal mobility through the contaminated soil.

Furthermore, McGowen et al. (in press) investigated the potential formation of metal phosphate precipitates or minerals formed from immobilization treatments by constructing activity-ratio diagrams and plotting chemical speciation data on the diagrams. The activity-ratio diagrams suggested that octavite ($CdCO_3$ log K_{SP} = −12.8) may control Cd solubility in the untreated soil with no added phosphorus. However, when phosphorus was added as DAP, cadmium phosphate $Cd_3(PO_4)_2$ (log K_{SP} = −38.1) was the potential mineral controlling Cd solubility (McGowen et al., in press). This result indicated that DAP application shifted the mineral-controlled solubility of Cd from a relatively soluble Cd-carbonate (octavite) to a sparingly soluble Cd-phosphate. For Pb, anglesite ($PbSO_4$ log K_{SP} = −7.79) was indicated as the mineral-controlling Pb solubility in soil without added phosphate. With the addition of phosphate as DAP, activity-ratio diagrams suggested that hydroxypyromorphite (log K_{SP} = −76.8) becomes the mineral-controlling Pb solubility. Similar to the results obtained for Cd, this suggests that DAP shifted the mineral controlled solubility from a relatively soluble $PbSO_4$ to the sparingly soluble Pb-hydroxypyromorphite. Activity-ratio diagrams for Zn did not indicate a shift in the mineral-controlled solubility, with solution Zn being controlled by hopeite ($Zn_3(PO_4)_2 \cdot 4H_2O$) or Zn-pyromorphite in soils with or without added phosphate.

Reductions in metal solubility or concentration of soluble metal species from chemical amendments through increased sorption or precipitation of insoluble metal minerals are effective means of decreasing metal transport from smelting wastes, mine tailings, and contaminated soils.

Table 4.2 Soil Solution Cd, Pb, and Zn Speciation Data from Untreated and DAP-Amended Soils

Metal	P Treatment (mg kg⁻¹)	Metal Species (% of total metal in solution)					Σ % Low Mobility[a]	Σ % High Mobility[b]	Total Conc. (mg L⁻¹)	High Mobility Conc. (mg L⁻¹)
		M^{2+}	MOH^+	MSO_4^0	$M(SO_4)_2^{2-}$	Other				
Cd	0	54.6	—	38.0	5.5	1.9	54.6	43.5	7.2	3.1
	460	35.5	—	43.3	20.9	0.3	35.5	64.2	3.7	2.3
	920	39.1	—	43.0	17.1	0.8	39.1	60.1	2.4	1.5
	2300	22.4	—	36.5	38.8	2.3	22.4	75.3	1.0	0.78
Pb	0	44.5	4.8	45.4	4.1	1.2	49.3	49.5	0.20	0.10
	460	29.0	2.5	52.3	15.9	0.3	31.5	68.2	0.17	0.12
	920	30.1	8.4	48.9	12.3	0.3	38.5	61.2	0.11	0.06
	2300	18.9	3.5	45.5	30.6	1.5	22.4	76.1	0.04	0.03
Zn	0	58.6	6.5	30.9	3.6	0.4	65.1	34.5	58	20
	460	42.1	3.7	39.0	14.9	0.3	45.8	53.9	4.0	2.1
	920	41.7	12.0	34.8	11.0	0.5	53.7	45.8	4.3	2.0
	2300	27.9	5.2	34.5	29.2	3.2	33.1	63.7	4.4	2.0

[a] Σ % Low mobility includes cationic metal species (M^{1+} M^{2+})
[b] Σ % High mobility includes uncharged and anionic metal species (M^0 M^{1-} M^{2-})

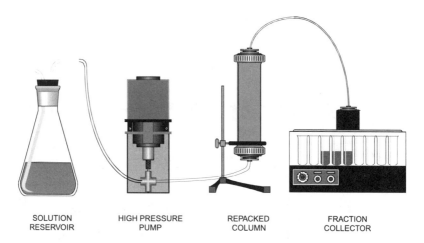

SOLUTION   HIGH PRESSURE   REPACKED   FRACTION
RESERVOIR   PUMP   COLUMN   COLLECTOR

Figure 4.1   Schematic of solute transport apparatus for evaluating chemical immobilization treatments to reduce Cd, Pb, and Zn solubility and transport in smelter contaminated soil.

Reduction in Metal Transport from Chemical Treatments

Most studies investigating chemical immobilization treatments have focused on reducing bioavailability (i.e., plants, gastrointestinal), solubility, or extractability (i.e., sequential extractions). Results from these studies may show reduced solubility and potential decreased metal transport, but few studies have measured reductions in metal transport from chemical amendments to contaminated soils.

Jones et al. (1997) investigated the transport of As in contaminated mine tailings following liming using column leaching experiments, chemical fractionation, and SEM-EDX. Their investigations showed that concentrations of soluble As in the tailings were highly correlated with pH. With two of the low pH experiments, lime was added and the material was leached under unsaturated conditions. Liming the mine tailings increased As concentration in column effluent 2 orders of magnitude over unlimed. The sequential extraction scheme and spectroscopic analyses revealed that increased mobility of As with liming was due to pH-dependent sorption on oxide minerals and not dissolution of arsenic mineral phases.

McGowen (2000) evaluated agricultural limestone, mineral rock phosphate, and diammonium phosphate chemical immobilization treatments to reduce Cd, Pb, and Zn solubility and transport in a smelter-contaminated soil. Chemical immobilization treatments were evaluated using solute transport experiments with repacked soil columns using modified methods similar to those described by Selim and Amacher (1997) (Figure 4.1). Plotted elution curves for Cd (Figure 4.2), Pb (Figure 4.3), and Zn (Figure 4.4) illustrate the differences in release and transport of heavy metals for each of the amendments. Furthermore, the numerical integration of observed metals in solution that were eluted through 60 pore volumes shows a direct comparison of total metal mass transported from each treatment (Table 4.3). Of the chemical treatments investigated, mixed rock phosphate treatments were the least effective for

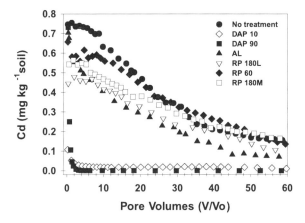

Figure 4.2   Cadmium elution curves for untreated and amended soils.

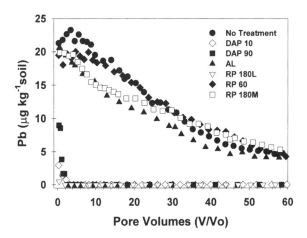

Figure 4.3   Lead elution curves for untreated and amended soils.

reducing total Cd, Pb, and Zn elution from the contaminated soil. Limestone, although moderately effective for reducing Cd and Pb (45% and 54.8% reduction over the untreated check, respectively), was less effective for decreasing Zn in effluent solution. Poor efficiency of the limestone treatment was ascribed to the soil system being prebuffered at a near neutral pH, and the slight increase in pH creating modest effects on metal solubility in the contaminated soil. Rock phosphate layered at 180 g kg^{-1} was the most effective for reducing Pb transport with a 99.9% reduction over the untreated control. The improved efficiency of the layered RP treatment over mixed treatments was attributed to greater effective surface area that allowed for immobilization of Pb within the rock phosphate matrix. However, DAP was superior to all other treatments for reducing Cd and Zn elution. The 10 g DAP kg^{-1} treatment was the most effective for immobilizing heavy metals transported from the contaminated soil, with only 5.4% of Cd, 1.1% of Pb, and 4.2% of the Zn eluted compared

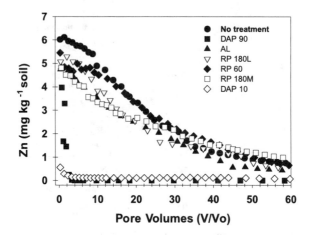

Figure 4.4   Zinc elution curves for untreated and amended soils.

Table 4.3   Cumulative Mass of As, Cd, Pb, and Zn Collected
from Miscible Displacement Experiments
through 60 Pore Volumes of Elution

Treatment	As (mg kg⁻¹)	Cd (mg kg⁻¹)	Pb (μg kg⁻¹)	Zn (mg kg⁻¹)
Control	ND	14.9	460	108
DAP 10	0.13	0.8	5.2	4.5
DAP 90	29.5	1.2	44.7	24.7
AL	0	6.7	252	84.4
RP 180L	0	7.0	0.5	81.6
RP 60	0	12.7	418	94.1
RP 180M	0	10.9	376	79.3

with the untreated soil. Increasing DAP from 10 to 90 g kg⁻¹ released As via ligand exchange and elevated total As eluted from 0.13 to 29.5 mg kg⁻¹. Peryea and Kammereck (1997) reported similar phosphate-arsenate ligand exchange reaction when investigating the release and movement of As from additions of phosphate fertilizers to arsenate-contaminated orchard soils.

In addition to the interpretations provided by measuring metal elution from chemically amended contaminated soil, McGowen et al. (in press) also applied solute transport models to quantitatively describe the effects of specific chemical treatments on the release and transport of heavy metals. In this study, diammonium phosphate was evaluated at three different rates (460, 920, and 2300 mg P kg⁻¹) to determine the optimum amount for reducing Cd, Pb, and Zn elution from a smelter-contaminated soil. Observed metal elution curves were fitted with a transport model for quantitative comparisons of retardation (R) and distribution coefficients (K_d). Model-fitted elution curves for Cd, Pb, and Zn are shown as solid lines in Figures 4.5, 4.6, and 4.7. Best-fit retardation (R), and calculated distribution coefficients (K_d) from these experiments are given in Table 4.4. Fitted metal elution curves for most treatments

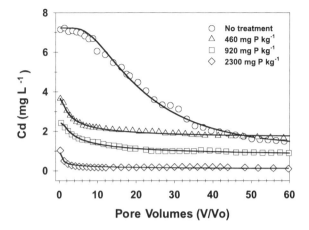

Figure 4.5   Observed (symbol) and fitted (line) cadmium elution curves for untreated and diammonium phosphate amended contaminated soils.

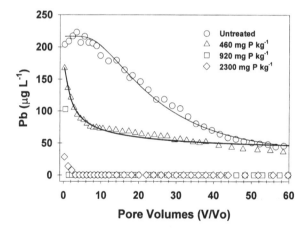

Figure 4.6   Observed (symbol) and fitted (line) lead elution curves for untreated and diammonium phosphate amended contaminated soils.

had $r^2 > 0.9$ and were well described by the COLUMN model. Model-fitted metal elution curves showed increasing retardation (R) with increasing P application (Table 4.4). Retardation factors increased approximately twofold for Cd and sixfold for Zn between the untreated soil and the 2300 mg P kg^{-1} treatments. Lead retardation factors increased approximately 3.5-fold with the addition of 460 mg P kg^{-1}. These increases in R with added DAP indicate that metal breakthrough is slowed by increasing P applications. In addition to increased retardation factors (R) determined from fitted metal elution curves, DAP increased distribution coefficients (K_d) with corresponding increases in phosphorus application (Table 4.4). Distribution coefficients calculated for each metal indicate slower rates of metal partitioning from sorbed/precipitated phases to mobile phases, with rates that decrease with corresponding increases

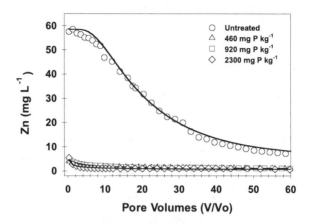

Figure 4.7 Observed (symbol) and fitted (line) zinc elution curves for untreated and diammonium phosphate amended contaminated soils.

**Table 4.4 Summary of Transport Parameters, Best-Fit Retardation (R), and Calculated
Distribution Coefficients (K_d) for Cd, Pb, and Zn Elution from Untreated and
DAP-Amended Soils**

P Treatment (mg P kg⁻¹)	Cd			Pb			Zn		
	R	K_d (L kg⁻¹)	r^2	R	K_d (L kg⁻¹)	r^2	R	K_d (L kg⁻¹)	r^2
0	14.8	4.0	0.995	10.9	2.9	0.992	9.5	2.5	0.993
460	16.7	4.7	0.987	37.7	10.8	0.970	15.9	4.5	0.984
920	18.1	4.9	0.993	58.2[a]	16.4	—	20.9	5.7	0.992
2300	31.4	9.0	0.923	30.6[a]	8.74	—	58.8	17.1	0.903

[a] Estimated from equation based on practical quantitative limit (PQL) of instrument method.

in DAP treatment. Calculated distribution coefficients increased from 4.0 to 9.0 L
kg⁻¹ for Cd, from 2.9 to 10.8 L kg⁻¹ for Pb, and from 2.5 to 17.1 L kg⁻¹ for Zn.
Increased K_d values were likely due to the formation of metal phosphate precipitates
and/or increased sorption. As indicated by the retardation and distribution coefficients from the model-fitted elution curves, the 2300 mg P kg⁻¹ application was the
most effective for immobilizing Cd, Pb, and Zn eluted from the contaminated soil
when compared with the untreated check.

SUMMARY

There are many methods for describing and interpreting metal release and transport from mine- and smelter-contaminated soils. Combinations of the methods
described may offer the best possible solution to understanding metal release and
transport from the plethora of mining and smelting sites throughout the world. Often
the heterogeneous nature of contaminants, soil, weathering conditions, and other

modifying factors will diminish power of assessment using a single-tiered research approach. In these cases, a multifaceted research approach is warranted.

New combinations and modifications of traditional methods of assessing metal release and transport in contaminated soils hold a great deal of promise for the investigation of chemical methods to reduce metal release and transport from soils contaminated by mining and smelting activities. Future research activity should endeavor to provide quantitative scientific information for understanding the chemical processes and mechanisms of metal release and the interactions between metal species in solution and the soil during transport processes.

REFERENCES

Abdel-Saheb, I., A.P. Schwab, M.K. Banks, and B.A. Hetrick. 1994. Chemical characterization of heavy-metal contaminated soil in southeast Kansas. *Water Air Soil Pollut.* 78:73-82.

Basta, N.T. and R. Gradwohl. 2000. Estimation of Cd, Pb, and Zn bioavailability in smelter-contaminated soils by a sequential extraction procedure. *J. Soil Contam.* 9:149-164.

Benner, S.G., E.W. Smart, and J.N. Moore. 1995. Metal behavior during surface-groundwater interaction, Silver Bow Creek, Montana. *Environ. Sci. Technol.* 29:1789-1795.

Cernik, M., P. Federer, M. Borkovec, and H. Sticher. 1994. Modeling of heavy metal transport in a contaminated soil. *J. Environ. Qual.* 23:1239-1248.

Chen, X., J.V. Wright, J.L. Conca, and L.M. Peurrung. 1997. Evaluation of heavy metal remediation using mineral apatite. *Water Air Soil Pollut.* 98:57-78.

Chlopecka, A. and D.C. Adriano. 1996. Mimicked *in situ* stabilization of metals in a cropped soil: bioavailability and chemical form of zinc. *Environ. Sci. Technol.* 30:3294-3303.

Cooper, E.M., D.G. Strawn, J.T. Sims, D.L. Sparks, and B.M. Onken. 1998. Effect of chemical stabilization by phosphate amendment on the desorption of P and Pb from a contaminated soil. p. 343. In *1998 Agronomy Abstracts,* ASA, Madison, WI.

Cotter-Howells, J. and S. Capron. 1996. Remediation of contaminated land by formation of heavy metal phosphates. *Appl. Geochem.* 11:335-342.

Dowdy, R.H. and V.V. Volk. 1983. Movement of heavy metals in soils. In *Chemical Mobility and Reactivity in Soils Systems,* SSSA Special Pub. No. 11, Soil Sci. Soc. Am., Madison, WI.

Dudka, S. and D.C. Adriano. 1997. Environmental impacts of metal ore mining and processing: a review. *J. Environ. Qual.* 26:590-602.

Eighmy, T.T., B.S. Crannell, L.G. Butler, F.K. Cartledge, E.F. Emery, D. Oblas, J.E. Krzanowski, J.D. Eusden, Jr., E.L. Shaw, and C.A. Francis. 1997. Heavy metal stabilization in municipal solid waste combustion dry scrubber residue using soluble phosphate. *Environ. Sci. Technol.* 31:3330-3338.

Fanfani, L., P. Zuddas, and A. Chessa. 1997. Heavy metal speciation analysis as a tool for studying mine tailings weathering. *J. Geochem. Explor.* 58:241-248.

Filius, A., T. Streck, and J. Richter. 1998. Cadmium sorption and desorption in limed topsoils as influenced by pH: isotherms and simulated leaching. *J Environ. Qual.* 27:12-18.

Fuge, R., F.M. Pearce, N.J.G. Pearce, and W.T. Perkins. 1993. Geochemistry of Cd in the secondary environment near abandoned metalliferous mines, Wales. *Appl. Geochem., Suppl. Iss.* 2:29-35.

Gee, C., M.H. Ramsey, J. Maskall, and I. Thornton. 1997. Mineralogy and weathering processes in historical smelting slags and their effect on the mobilisation of lead. *J. Geochem. Explor.* 58:249-257.

Hagni, R.D. and A.M. Hagni. 1991. Determination of the mineralogy of dusts near a Missouri lead sinter plant and blast furnace by reflected light and scanning electron microscopy. Minerals, Metals, and Materials Society (TMS), annual meeting, New Orleans, Feb. 24-28. V.11 67-73.

Hettiarachchi, G.M., G.M. Pierzynski, J. Zwonitzer, and M. Lambert. 1997. Phosphorus source and rate effects on cadmium, lead, and zinc bioavailabilities in a metal-contaminated soil. p. 463-464. In *Extended Abstr.,* 4th Int. Conf. on the Biogeochem. Trace Elements (ICOBTE), Berkeley, CA, 23-26 June, 1997.

Jones, C.A., W.P. Inskeep, and D.R. Neuman. 1997. Arsenic transport in contaminated mine tailings following liming. *J. Environ. Qual.* 26:433-439.

Jurinak, J.J. and J. Santillian-Medrano. 1974. The chemistry and transport of lead and cadmium in soils. Report 18, Utah Agric. Exp. Sta. Logan, UT.

Lambert, M., G. Pierzynski, L. Erickson, and J. Schnoor. 1994. Remediation of lead, zinc, and cadmium contaminated soils. *Issues Environ. Sci. Technol.* 7:91-102.

Laperche, V., S.J. Traina, P. Gaddam, and T.J. Logan. 1997. Chemical and mineralogical characterizations of Pb in a contaminated soil: reactions with synthetic apatite. *Environ. Sci. Technol.* 30:3321-3326.

Levi-Minzi, R. and G. Petruzzelli. 1984. The influence of phosphate fertilizers on Cd solubility in soil. *Water Air Soil Pollut.* 23:423-429.

Li, Z. and L.M. Shuman. 1996. Heavy metal movement in metal-contaminated soil profiles. *Soil Sci.* 161:656-666.

Logan, T.J. 1992. Reclamation of chemically degraded soils. *Adv. Soil Sci.* 17:13-35.

Ma, Q.Y., T.J. Logan, and S.J. Traina. 1995. Lead immobilization from aqueous solutions and contaminated soils using phosphate rocks. *Environ. Sci. Technol.* 29:1118-1126.

Ma, Q.Y., S.J. Traina, and T.J. Logan. 1993. *In situ* lead immobilization by apatite. *Environ. Sci. Technol.* 27:1803-1810.

Ma, L.Q. and G.N. Rao. 1997. Effects of phosphate rock on sequential chemical extraction of lead in contaminated soils. *J. Environ. Qual.* 26:788-794.

Mattigod, S.V., A.L. Page, and I. Thornton. 1986. Identification of some trace metal minerals in a mine-waste contaminated soil. *Soil Sci. Soc. Am. J.* 50:254-258.

Maskall, J., K. Whitehead, and I. Thornton. 1994. Migration of metals in soils and rocks at historical lead smelting areas. *Environ. Geochem. Health.* 16:2-82

McBride, M.B. 1989. Reactions controlling heavy metal solubility in soils. *Adv. Soil Sci.* 10:1-56.

McBride, M., S. Sauve, and W. Hendershot. 1997. Solubility control of Cu, Zn, Cd, and Pb in contaminated soils. *Eur. J. Soil Sci.* 48:337-346.

McGowen, S.L. 2000. *In situ* chemical treatments for reducing heavy metal solubility and transport in smelter contaminated soil. Ph.D. dissertation, Oklahoma State Univ., Stillwater, OK.

McGowen, S.L., N.T. Basta, and G.O. Brown. 2001. Use of diammonium phosphate to reduce heavy metal solubility and transport in smelter-contaminated soil. *J. Environ. Qual.* 30:493-500.

McLean, J.E. and B.E. Bledsoe. 1992. Behavior of metals in soils. EPA, Ground Water Issue, U.S. Environ. Protection Agency, EPA/540/S-92/018, U.S. Govt. Print. Office, Washington, D.C.

Mench, M., J. Vangronsveld, N.W. Lepp, and R. Edwards. 1998. Physico-chemical aspects and efficiency of trace element immobilization by soil amendments. In J. Vangronsveld and S.D. Cunningham (Eds.) *Metal-Contaminated Soils: In Situ Inactivation and Phytorestoration.* Springer-Verlag, Berlin.

Nordstrom, D.K. 1982. *Acid Sulfate Weathering.* SSSA Spec. Publ. 10:37-62.

Paulson, A.J. 1997. The transport and fate of Fe, Mn, Cu, Zn, Cd, Pb, and SO_4 in a groundwater plume and in downstream surface water in the Coeur d'Alene mining district, Idaho, U.S.A. *Appl. Geochem.* 12:447-464.

Peryea, F.J. and R. Kammereck. 1997. Phosphate-enhanced movement of arsenic out of lead arsenate-contaminated topsoil and through uncontaminated subsoil. *Water Air Soil Pollut.* 93:243-254.

Pierzynski, G.M. and A.P. Schwab. 1993. Bioavailability of zinc, cadmium, and lead in a metal-contaminated alluvial soil. *J. Environ. Qual.* 22:247-254.

Santillian-Medrano, J. and J.J. Jurinak. 1975. The chemistry of lead and cadmium in soil: solid phase formation. *Soil Sci. Soc. Am. Proc.* 39:851-856.

Scocart, P.O., K. Meeus-Verdinne, and R. DeBorger. 1983. Mobility of heavy metals in polluted soils near zinc smelters. *Water Air Soil Pollut.* 20:451-463.

Selim, H.M. 1992. Modeling the transport and retention of inorganics in soils. *Adv. Agron.* 47:331-384.

Selim, H.M. and M.C. Amacher. 1997. *Reactivity and Transport of Heavy Metals in Soils.* CRC Press, Boca Raton, FL.

Selim, H.M., M.C. Amacher, and I.K. Iskandar. 1990. Modeling the transport of heavy metals in soils. CRREL-Monograph 90-2, U.S. Army Corps of Engineers, Hanover, NH.

Shackelford, C.D. and M.J. Glade. 1997. Analytical mass leaching model for contaminated soil and soil stabilized waste. *Ground Water.* 35(2):233-242.

Vangronsveld, J. and S.D. Cunningham. 1998. Introduction to the concepts. In J. Vangronsveld and S.D. Cunningham, Eds., *Metal-Contaminated Soils: In Situ Inactivation and Phytorestoration.* Springer-Verlag, Berlin.

Wilkens, B.J. and P.G. Loch. 1997. Accumulation of cadmium and zinc from diffuse immission on acid sandy soils, as a function of soil composition. *Water Air Soil Pollut.* 95:1-16.

Xie, R.J. and A.F. MacKenzie. 1990. Sorbed ortho- and pyrophosphate effects on zinc reactions compared in three autoclaved soils. *Soil Sci. Soc. Am. J.* 54:744-750.

Zhang, P. and J.A. Ryan. 1999. Formation of chloropyromorphite from galena (PbS) in the presence of hydroxyapatite. *Environ. Sci. Technol.* 33:618-624.

Zhu, D., A.P. Schwab, and M.K. Banks. 1999. Heavy metal leaching from mine tailings as affected by plants. *J. Environ. Qual.* 28:1727-1732.

CHAPTER 5

Phase Plane Analysis and Dynamical System Approaches to the Study of Metal Sorption in Soils

Seth F. Oppenheimer, William L. Kingery, and Feng Xiang Han

INTRODUCTION

While considerable mathematical sophistication may be found in the soils liter-ature, there has not been much use made of the qualitative theory of differential equations. It is our intention in this chapter to do four things: (1) we will introduce the technique of phase plane analysis; (2) we will use this technique to address two problems from soil science, namely, the time it takes to reach equilibrium in a sorption problem and determining whether or not sorption is the only chemical process occurring in our experiments; (3) we will use a more qualitative approach to differential equations to develop a new model for multilayer sorption; and (4) we will consider hysteresis in desorption and develop and analyze an elementary model using phase plane analysis.

An elementary introduction to the mathematical techniques used here may be found in Blanchard et al. (1998) and Borrelli and Coleman (1998). A more sophis-ticated discussion may be found in Coddington and Levinson (1955) and Hirsch and Smale (1974). Other papers and abstracts where this sort of analysis is used in soil work are Kingery et al. (1998) and Oppenheimer et al. (in press). We use various numerical techniques in generating approximate solutions to systems of differential equations throughout this chapter; Burden and Faires (1997) will contain the needed background. Finally, we wrote our computer codes in the Matlab programming language (Math Works, Inc. 1992).

SINGLE DIFFERENTIAL EQUATIONS: PHASE LINE ANALYSIS

We want to use derivatives to describe physical phenomena, so we begin with a simple example that leads to a single ordinary differential equation.

Let us consider a chemical compound that is disappearing spontaneously from an aqueous solution with probability $P > 0$ in any given second. We assume we have a whole macroscopic sample with mass M grams. The mass of the sample will be proportional to the number of particles in the sample. Let us also assume we can measure the mass at time t seconds to get the mass measurement of $M(t)$. We expect that in a given second, the rate at which the mass is decaying is given by $pM(t)$ with units of gs^{-1} where $p = P \times$ grams per particle of the substance. Now we can also write the rate of change in the mass as dM/dt. Thus we end up with a differential equation:

$$\frac{dM}{dt} = -pM \qquad (5.1)$$

where the minus sign indicates disappearance.

Before attempting a solution to Equation 5.1 we can get information on the process by looking at a *phase line* for this equation. The phase line is simply a number line provided with arrows to indicate in what direction a point will move (Figure 5.1).

The point $M = 0$ is called a *stationary* or *equilibrium point*. If $M = 0$, there will be no change over time. Notice, that $M = 0$ is the value for M which makes $dM/dt = 0$. For $M > 0$, the arrow points down, because when we have a positive mass of material, the mass will decrease over time. The up arrow for $M < 0$ is a mathematical artifact because we cannot have negative mass. Mathematically, it states if there were such a thing as negative mass, it would tend to become less negative over time. By mathematical artifact here, we mean a fact about the mathematical model that does not relate to the physical system we are trying to model.

Notice that all of the arrows point toward $M = 0$; this means that 0 is an attracting equilibrium point or *sink*.

We can approach this problem analytically and obtain

$$M(t) = M_0 e^{-pt} \qquad (5.2)$$

where $M_0 = M(0)$. Notice that the analytic solution does exactly what the phase line diagram says it should; go to $M = 0$. If we could not find an analytic solution, we could do a numerical approximation, although we might lose our understanding of the long-term behavior of the system.

In another numerical example, let us assume we are adding mass at the rate of a gs^{-1}. Our rate of change is now the loss of $pM(t)$ plus a gs^{-1}. This gives us a differential equation of:

$$\frac{dM}{dt} = -pM + a \qquad (5.3)$$

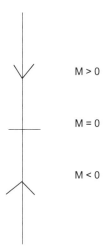

M > 0

M = 0

M < 0

Figure 5.1 Phase line diagram for Equation 5.1.

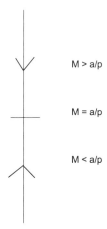

M > a/p

M = a/p

M < a/p

Figure 5.2 Phase line diagram for Equation 5.3.

What are our equilibrium points? We obtain these by solving the problem dM/dt = 0, or $-pM + a = 0$, or $M = a/p$. When, M > a/p, we will get dM/dt < 0, and when $M < a/p$, we will get $dM/dt > 0$ as is reflected in the phase line (Figure 5.2).

Here we have no physical problems with $M < a/p$ and we see that no matter what mass of material we start with, in the long run we tend toward an equilibrium of $M = a/p$. We note that we can draw the phase line in equilibrium mass of this manner because the "right-hand side" of Equation 5.3 does not depend explicitly on t, i.e., the equation is autonomous.

It happens that we can also find an analytic solution to the above problem using

$$M(t) = \frac{a}{p}\left(1 - e^{-pt}\right) + M_o e^{-pt} \tag{5.4}$$

but we have a great deal of information using only the phase line.

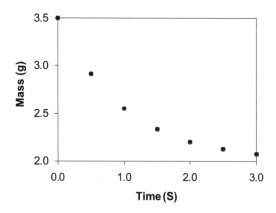

Figure 5.3   A plot of mass vs. time with a starting mass of $M_0 = 3.5$ for Equation 5.4.

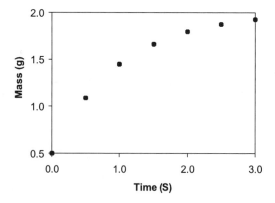

Figure 5.4   A plot of mass vs. time with a starting mass of $M_0 = 0.5$ for Equation 5.4.

It is worthwhile to consider how an experiment's data would normally be presented. For this purpose, we will take $a = 2$ and $p = 1$. In Figure 5.3, the graph is a plot of mass vs. time with a starting mass of $M_o = 3.5$.

In Figure 5.4, the graph is a plot of mass vs. time with a starting mass of $M_0 = 0.5$ for Equation 5.4. Looking at the two figures, we can see what is going on in terms of an approach to an equilibrium, but not as easily as when we use the phase line.

SYSTEMS OF TWO DIFFERENTIAL EQUATIONS AND PHASE PLANES

As in the case of modeling with a single differential equation, we shall discuss the techniques we wish to introduce in the context of a concrete example. In fact, our examples will reflect actual situations we have encountered in our research.

We consider an agitated vessel containing soil with a of total mass Mg and a solution of total volume VmL. We assume that there is a heavy metal that is dissolved in the solution that can be sorbed to the soil. The concentration of the metal in the solution at time t will be given by $c(t)$ μgmL^{-1} and the concentration of the metal sorbed to the soil will be given by $q(t)$ μg g^{-1}.

Recall that there exists an equilibrium relationship between the solution concentration and the sorbed concentration. That is, for any given solution concentration c_0 there is a sorbed concentration q_0 such that if the solution concentration is c_0 and the sorbed concentration is q_0 the concentrations will not change over time. This defines a functional relationship where one inputs the solution concentration and the output is the equilibrium sorbed concentration. The function, defined at a fixed temperature, is called the *sorption isotherm* and will be denoted by f. Some typical examples are the Henry isotherm:

$$q_0 = \gamma c_0 \tag{5.5}$$

the Langmuir isotherm:

$$q_0 = \frac{ac_0}{1+\beta c_0} \tag{5.6}$$

and the Freundlich isotherm:

$$q_0 = a(c_0)^\gamma \tag{5.7}$$

where a, β, and γ are positive constants. In this section, we will use a Langmuir isotherm

$$q_0 = \frac{100c_0}{1+.01c_0}$$

for our examples, where the constants have been chosen for convenience. This is plotted in Figure 5.5. (We note that we are ignoring the possibility of hysteresis effects that would allow for multiple equilibrium sorbed concentrations and possible differing sorption and desorption behaviors. We will discuss this in the section titled Desorption–Sorption Modeling.)

It is important to realize that understanding the equilibrium behavior is not sufficient for problems involving transport, such as site remediation and the study of agricultural wastes. Therefore, we must give a model for how c and q change in time. We will assume that the system will tend toward a condition of equilibrium and the rate of change will be proportional to the distance from equilibrium. That is to say, the rate at which the sorbed concentration is changing at time t will be proportional to $f(c(t)) - q(t)$. Notice that if the current sorbed concentration is smaller

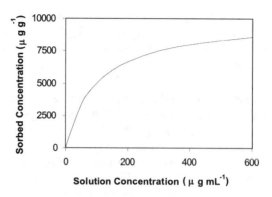

Figure 5.5   Langmuir isotherm of a chemical compound on a soil.

than the equilibrium sorbed concentration for the current solution concentration, $f(c(t))$, the rate of change will be positive. On the other hand, if the current sorbed concentration is larger than the equilibrium sorbed concentration for the current solution concentration, the rate of change will be negative. This leads to the equation

$$\frac{dq}{dt} = r_q\big(f(c(t)) - q(t)\big)$$
(5.8)

where the rate constant r_q is the constant of proportionality with units of s^{-1}. Similarly, we obtain an equation for the rate of change in the solution concentration

$$\frac{dc}{dt} = r_c\big(q(t) - f(c(t))\big)$$
(5.9)

where r_c has units of $g\ mL^{-1}\ s^{-1}$. This yields a two-by-two system of ordinary differential equations

$$\frac{dc}{dt} = rc\big(q(t) - f(c(t))\big)$$

$$\frac{dq}{dt} = r_q\big(f(c(t)) - q(t)\big)$$
(5.10)

Again we note that the "right-hand sides" of the equations have no explicit dependence on t. That is, t does not appear except as an argument of c and q, and the equations are autonomous.

We will now discuss some ways of analyzing this system without solving it or considering experimental data. We will then apply this analysis to two data sets. Let us again consider the graph of the isotherm, but now we will view it as a phase plane which we can use to understand what the dynamical behavior of the system will be.

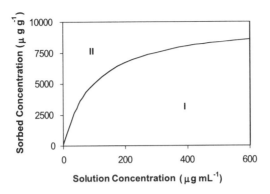

Figure 5.6 Langmuir isotherm as a phase plane with two distinct states, I and II.

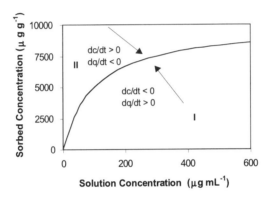

Figure 5.7 Langmuir isotherm with the motion of the state vector and the derivatives of rates of solutes in the solution and solid states.

As we see in Figures 5.6 and 5.7, the isotherm divides the first quadrant of the plane into two distinct regions, I and II. Each point on the plane represents a possible state of the system; the horizontal coordinate giving the solution concentration and the vertical coordinate giving the sorbed concentration. In region I we have $f(c) > q$ and thus from Equation 5.10 we have

$$\frac{dc}{dt} < 0 \quad \text{and} \quad \frac{dq}{dt} > 0 \tag{5.11}$$

In a similar fashion, for region II we have $f(c) < q$ and thus

$$\frac{dc}{dt} > 0 \quad \text{and} \quad \frac{dq}{dt} < 0 \tag{5.12}$$

This means that if, at a given time t the state of our system places it in region I, the solution concentration will be decreasing and the sorbed concentration will be increasing and the point that represents the state of the system will be moving up and to the left, toward an equilibrium point on the isotherm. Similarly, if, at a given time t, the state of our system places it in region II, the solution concentration will be increasing and the sorbed concentration will be decreasing and the point that represents the state of the system will be moving down and to the right toward an equilibrium point on the isotherm. This analysis tells us what behavior we can expect from our experimental data *if our model is correct* and the experiment has been done correctly.

Using conservation of mass, we can get still more information about how our experimental data can be expected to behave if our model is correct. At $t = 0$ we will know the initial concentrations $c(0)$ and $q(0)$. Since no mass is removed from the system and since mass is neither created nor destroyed, the total mass at any given t should remain constant. That is, the total initial mass of contaminant equal to the mass in solution plus mass sorbed to the soil $= c(0)V + q(0)M$ must be the same at any given t, or

$$c(t)V + q(t)M = c(0)V + q(0)M \qquad (5.13)$$

Thus, all of our states $(c\,(t), q\,(t))$ should lie on the line

$$q(t) = -\frac{V}{M}c(t) + \frac{V}{M}c(0) + q(0) \qquad (5.14)$$

In fact, it is exactly this relationship we use to obtain the sorbed concentration from the solution concentration in experiments. We will see an example later where this breaks down.

The same sort of conservation of mass analysis forces a relationship between r_q and r_c. The rate at which the mass of metal disappears from the solution should equal the rate at which mass is sorbed onto the soil. That is,

$$-\frac{dc}{dt}V = \text{rate at which mass is removed from the solution}$$

$$\qquad (5.15)$$

$$= \text{rate at which mass sorbs onto soil} = \frac{dq}{dt}M$$

This implies that

$$-r_c\big(q(t) - f(c(t))\big)V = r_q\big(f(c(t)) - q(t)\big)M \qquad (5.16)$$

or

$$\frac{r_c}{r_q} = \frac{M}{V} \qquad (5.17)$$

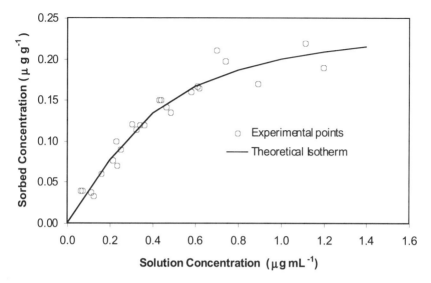

Figure 5.8   An example of Langmuir-Freundlich isotherm of a protein sorption on calcium-saturated Wyoming smectite.

(It is a useful exercise for the reader to make sure that the units work out correctly.) This can be used as a check when we are seeking to find the rate constants from experimental data.

TWO EXAMPLES

We now give the results of two sets of experimental data. We will be working from data from experiments studying the sorption of calcium to a Wyoming clay. We will provide a brief note on experimental means and methods at the end of this chapter.

We first employed an equilibrium isotherm in the form of a Langmuir-Freundlich curve

$$f(c) = \frac{\alpha c^{\gamma}}{1 + \beta c^{\gamma}} \qquad (5.18)$$

We identified the parameters α, β, and γ to obtain $\alpha = 1.057$, $\beta = 4.298$, and $\gamma = 1.396$ minimizing a sum of square error objective function, constructed using experimental data. The fitted curve and the experimental values are compared in Figure 5.8.

How long should an experiment measuring the dynamic process of sorption last? Recall that we are assuming the differential equation model (Equation 5.10) with the Langmuir-Freundlich isotherm given above. After two hours, we recorded our data points on the phase plane as seen in Figure 5.9. Clearly, two hours are not enough to reach equilibrium. Merely plotting our data in phase space showed us

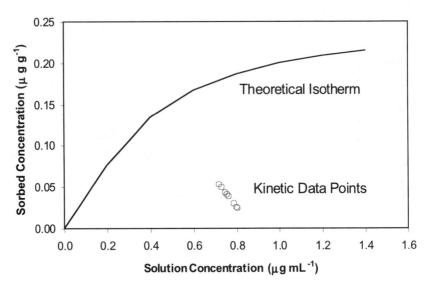

Figure 5.9 Langmuir-Freundlich isotherm and experimental kinetics during the first 120 min of sorption of a protein on calcium-saturated Wyoming smectite.

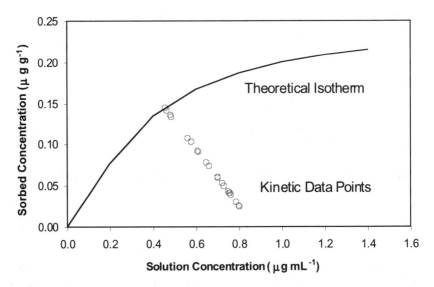

Figure 5.10 Langmuir-Freundlich isotherm and experimental kinetics during 2880 min of sorption of a protein on calcium-saturated Wyoming smectite.

that we need to run our kinetic experiments longer. Our results for 2880 minutes are shown in Figure 5.10.

We can also plot the predicted path in phase space predicted by Equation 5.10 with $r_c = .0032$ and $r_q = .0011$ as shown in Figure 5.11. In this case, $M/V = 2.9$ and $r_c/r_q = 2.91$.

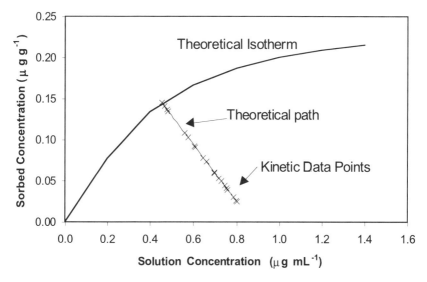

Figure 5.11 Langmuir-Freundlich isotherm, theoretical and experimental kinetics during 2880 min of sorption of a protein on calcium-saturated wyoming smectite.

Our next example uses the same identified isotherm. We ran some experiments with somewhat higher initial solution concentrations. Our results, plotted on the phase plane, are given below. We note that the rate constants identified using this data set should be the same as those we identified earlier; however, they are not, and we have $r_c = .0054$ and $r_q = .0016$. In this case, $M/V = 2.9$ and $r_c/r_q = 3.38$. This leads to a loss of conservation of mass in the theoretical results.

In Figure 5.12 we see a problem. Notice that the data points start at the bottom of the quadrant and move through and above the isotherm. Now the isotherm represents a continuous set of equilibrium points to the set of differential equations, (5.10), we are using to model the experiment. This means that our data passes through an equilibrium point. Mathematically, this is impossible. Therefore, either there is a problem with our data or there is a problem with our model. It is possible that the experiment to determine the equilibrium concentration was not run long enough to reach equilibrium, which would result in an incorrectly identified isotherm. However, this isotherm worked well at the lower concentration. Among the possible difficulties for our model are: (1) perhaps the isotherm is not accurate for high concentrations of contaminant; (2) perhaps there are other mechanisms coming into play that only occur at high concentrations such as precipitation (Oppenheimer et al., in press); or, multisite sorption. In any case, the phase plane analysis of the data has shown us that there is a problem with our model or our experiment.

A NEW MULTILAYER SORPTION MODEL

The following new model for multilayer sorption is worked out in detail for a two-layer model. However, the same reasoning can be used for any fixed number

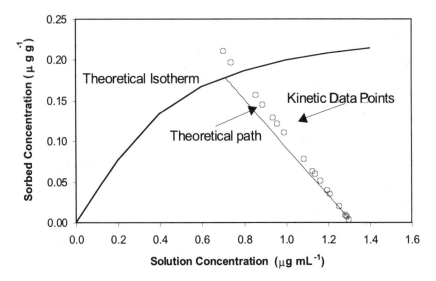

Figure 5.12 Langmuir-Freundlich isotherm, theoretical and experimental kinetics during 2880 of sorption of a protein on calcium-saturated Wyoming smectite at the high initial protein concentrations.

of layers. Let us consider two possible ways a metal can sorb to a soil. The first way is to directly bond to the soil, and the second way is to bond to metal bound to the soil.

Let us define four quantities:

1. The concentration of mass of metal in solution per unit volume of fluid at t, $c(t)$
2. The concentration of mass of metal bound to the soil per unit mass of soil at t, $q(t)$
3. The concentration of mass of metal bound to the metal bound to the soil per unit mass of metal bound to the soil at t, $D(t)$
4. The concentration of mass of metal bound to the metal bound to the soil per unit mass of soil at t, $\delta(t)$

We assume the standard equilibrium relationship of $q = f(c)$ where f is the functional representation of the equilibrium isotherm. We will also assume that we have an equilibrium relationship of $D = g(c)$. Let us ask what physical relationships we have. We will assume that there is a volume V of fluid and a mass M of soil in a well-mixed container. We can expect the rates of change in metal concentration to follow

$$V\frac{dc}{dt} = -M\left(\frac{dq}{dt} + \frac{d\delta}{dt}\right) \tag{5.19}$$

or

$$\frac{dc}{dt} = \frac{M}{V}\left(\frac{dq}{dt} + \frac{d\delta}{dt}\right) \tag{5.20}$$

This will only hold for batch sorption experiments. However, we can use this to find our local kinetics. The rate of change in q is assumed to be first order, that is to say, proportional to the difference between $q(t)$ and its equilibrium value:

$$\frac{dq}{dt} = r_{qc}\left(f(c(t)) = -q(t)\right) \tag{5.21}$$

We have the same first-order relationship for $D(t)$:

$$\frac{dD}{dt} = r_{Dc}\left(g(c(t)) - D(t)\right) \tag{5.22}$$

We may now observe that $D = \delta\, q^{-1}$ and rewrite our equations using this. We do the details for the third equation:

$$\begin{aligned}
r_{Dc}\left(g(c(t)) - D(t)\right) &= \frac{dD}{dt} \\[6pt]
&= \frac{d}{dt}\,\delta(t)q(t)^{-1} \\[6pt]
&= \frac{d\delta}{dt}\,q(t)^{-1} - q(t)^{-2}\delta(t)\frac{dq}{dt} \\[6pt]
&= \frac{d\delta}{dt}\,q(t)^{-1} - q(t)^{-2}\delta(t)r_{qc}\left(f(c(t)) - q(t)\right)
\end{aligned} \tag{5.23}$$

We now solve for $d\delta/dt$ to obtain

$$\frac{d\delta}{dt} = q(t)^{-1}\delta(t)r_{qc}\left(f(c(t)) - q(t)\right) + q(t)r_{Dc}\left(g(c(t)) - \delta(t)q(t)^{-1}\right) \tag{5.24}$$

We also need to rewrite our equation for dc/dt:

$$\begin{aligned}
\frac{dc}{dt} &= -\frac{M}{V}\left(\frac{dq}{dt} + \frac{d\delta}{dt}\right) \tag{5.25} \\[6pt]
&= -\frac{M}{V}\Big[r_{qc}\left(f(c(t)) - q(t)\right) + q(t)^{-1}\delta(t)r_{qc}\left(f(c(t)) - q(t)\right) \\[6pt]
&\qquad\qquad + q(t)r_{Dc}\left(g(c(t)) - \delta(t)q(t)^{-1}\right)\Big] \\[6pt]
&= r_{cq}\left(q(t) - f(c(t))\right)\left(1 + \delta(t)q(t)^{-1}\right) + q(t)r_{cD}\left(\delta(t)q(t)^{-1} - g(c(t))\right)
\end{aligned}$$

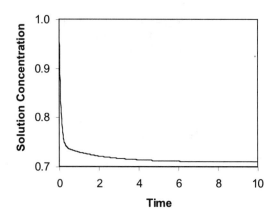

Figure 5.13 Theoretical plots of the metal concentration in solution with time.

where $r_{cq} = (M/V) \, r_{qc}$ and $r_{cD} = (M/V) \, r_{Dc}$. This argument by conservation of mass holds only in our batch situation. However, the constants and equations we obtain will hold for transport problems. We now replace D with δq^{-1} in Equations 5.23 through 5.25 to obtain

$$\frac{dc}{ct} = r_{cq}\big(q(t) - f(c(t))\big)\big(1 + \delta(t)q(t)^{-1}\big) + q(t)r_{cD}\big(\delta(t)q(t)^{-1} - g(c(t))\big) \quad (5.26)$$

$$\frac{dq}{dt} = r_{qc}\big(f(c(t)) - q(t)\big)$$

$$\frac{d\delta}{dt} = q(t)^{-1}\delta(t)r_{qc}\big(f(c(t)) - q(t)\big) + q(t)r_{Dc}\big(g(c(t)) - \delta(t)q(t)^{-1}\big)$$

This is noticeably more complex than the standard double isotherm model (Selim and Amacher, 1997).

We will now do some numerical experiments. In the example below, $r_{cq} = r_{qc} = 10$ and $r_{cD} = r_{Dc} = .5$. We use Langmuir sorption

$$f(c) = \frac{.6c}{1+c}$$

$$g(c) = \frac{.4c}{1+c}$$

Our initial conditions are $c(0) = 1$, $q(0) = \delta(0) = 0$. Our first time plot for concentration is seen in Figure 5.13. Now we show $q + \delta$ vs. time in Figure 5.14. In fact, this is all we know about how to measure at this point. We show the phase diagram in Figure 5.15. Finally, we will plot the two sorbed concentrations against time on the same plot in Figure 5.16.

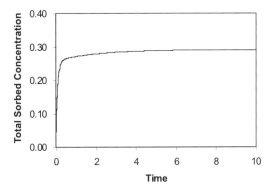

Figure 5.14   Theoretical plots of the metal concentration sorbed on solid (as the sum of the first and second layers) with time.

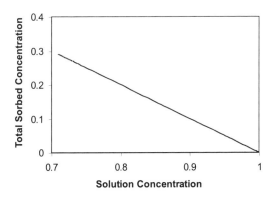

Figure 5.15   Phase plane diagram of the total metal concentration sorbed on the solid against in solution.

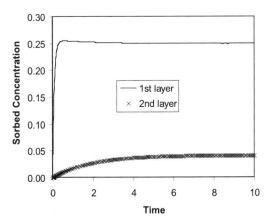

Figure 5.16   Plots of the metal sorbed in the first and second layers on the solid state with time.

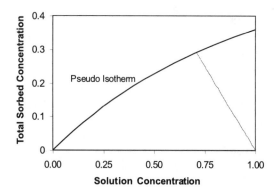

Figure 5.17 Phase plane diagram of the metal in solution and sorbed on the solid (including both isotherm and kinetics).

We can very clearly see the different rates come into play here. We might even be seeing some first-layer desorption. We can get what the isotherm would look like by setting all of the derivatives to 0:

$$\delta + q = \big(1 + g(c)\big)f(c)$$

We plot this along with the phase plane in Figure 5.17.

DESORPTION-SORPTION MODELING

The understanding of sorption/desorption phenomena is not just of scientific interest. Efficient site remediation will be impossible without a thorough understanding of sorption/desorption. A fundamental mistake is sometimes made in assuming that sorption is a reversible process both with respect to equilibrium behavior and dynamic behavior. In fact, there is considerable evidence indicating that sorption is a hysteresis process. That is, behavior during sorption and desorption differs. A different approach to sorption hysteresis may be found in Showalter and Peszynska (1998).

We can see at least two different types of sorption hysteresis, equilibrium hysteresis and dynamic hysteresis. We will describe the problem in terms of zinc sorbing to a clay, which is the example we will give experimental data for.

1. **Equilibrium hysteresis**. If the behavior of sorption and desorption differ in equilibria, we expect to see different equilibria reached depending on whether the system is sorbing or desorbing. Consider Figure 5.18. The horizontal axis is for c, the concentration of zinc in solution in $\mu g/mL$, and the vertical axis is for q, the concentration of zinc sorbed to the clay in $\mu g/g$. The lower curve is the sorption equilibrium isotherm and given by $q = f_s(c)$, and the upper curve is the desorption equilibrium isotherm and is given by $q = f_d(c)$. If hysteresis is occurring, a system starting with a state of (c_0, q_0) in region I will reach an equilibrium state $(\bar{c}, \bar{q})$

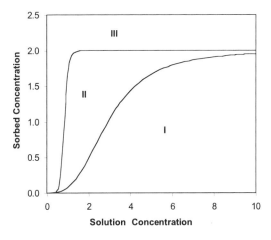

Figure 5.18 Hysteresis phase plane of the metal sorption/desorption in soils.

satisfying $\overline{q} = f_s(\overline{c})$. If hysteresis is occurring, a system starting with a state of (c_0, q_0) in region III will reach an equilibrium state $(\overline{c}, \overline{q})$ satisfying $\overline{q} = f_d(\overline{c})$. If hysteresis is occurring, a system starting with a state of (c_0, q_0) in region II is already in equilibrium.

2. **Kinetic hysteresis.** For a system not in equilibrium, the concentrations change with time. If $c(t)$ represents the solution concentration at time t, and $q(t)$ represents the sorbed concentration at time t, we assume that their rates of change satisfy

$$\begin{cases} \dfrac{dc}{dt} = -r_{cs}\left(q - f_s(c)\right)^- + r_{cd}\left(q - f_d(c)\right)^+ \\[2mm] \dfrac{dq}{dt} = r_{qs}\left(q - f_s(c)\right)^- - r_{qd}\left(q - f_d(c)\right)^+ \end{cases}$$

In an experimental situation, where the total mass is known, we will be able to reduce this to a single equation. In any case, even if $f_s = f_d$, we may still see a difference in the rate of change. That is, it may still be the case that we have a purely kinetic hysteresis in the sense that $r_s \neq r_d$. The region II along with the two boundary curves, is the set of equilibrium points for the dynamical system defined by the differential equations.

We will now examine four sets of experimental data, one for each sorption and desorption equilibrium and one each for a kinetic run for sorption and desorption.

Equilibrium

We assume that each process follows a Langmuir-Freundlich equilibrium. That is, for a fixed solution concentration, c, there is a fixed sorbed concentration q given by

$$q = f(c) = \frac{ac^\gamma}{1 + \beta c^\gamma}$$

where α, β, γ and are fixed parameters. That is, as above, the function for equilibrium may be different for sorption and desorption. For our data, this is the case. If f_s is the function giving the sorption isotherm and f_d is the function giving the desorption isotherm, we expect

$$f_s(c) \le f_d(c)$$

In general, if the solution concentration is a fixed value, say c_0, and $\bar{q}$ is any sorbed concentration satisfying

$$f_s(c_0) \le \bar{q} \le f_d(c_0)$$

we expect the system to be at rest. That is, the concentrations will not change over time.

Below are the identified parameters for sorption and desorption isotherms for zinc on a clay from Mississippi. We make no claims on the significance of all of the digits.

	α	β	γ
Miss. zinc sorption	256.7192	.0073	1.4343
Miss. zinc desorption	2591.4	.0913	2.4883

We give plots of the fitted curves along with the data used in deriving them in Figures 5.19 and 5.20. We note that our concentration scale for the desorption data is much narrower

Kinetic Model

Working with the identified isotherm, we calibrated the kinetic model for both sorption and desorption. We see that there were clearly difficulties. It is unclear whether these difficulties result from inadequacies in the model or errors in experimental measurements; the ratio

$$\frac{r_{qs}}{r_{cs}}$$

should equal

$$\frac{Solution\ volume}{Clay\ mass} = \frac{10ml}{.05g} = 500 \frac{ml}{g}$$

The deviation from 500 of this ratio is also an indication of difficulty maintaining conservation of mass. The table below contains the identified rate constants for sorption. We also include the starting mass and the finishing mass predicted by the calibrated model.

	r_{cs}	r_{qs}	r_{qs}/r_{cs}	Starting Mass (μg)	Ending Mass (μg)
Zinc on Mississippi clay	.0025	.8179	327.2	221.926	173.0812

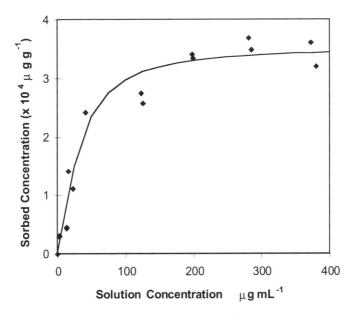

Figure 5.19   Zinc sorption isotherm on a Mississippi smectite.

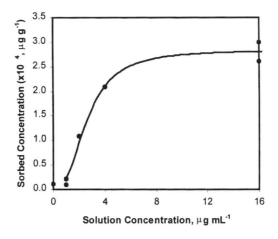

Figure 5.20   Zinc desorption isotherm on a Mississippi smectite.

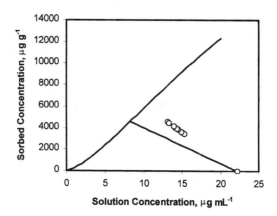

Figure 5.21 Zinc sorption isotherm, theoretical and experimental sorption kinetics on a Mississippi smectite.

We provide a phase plane diagram with the sorption isotherm, the kinetic run data, and the predicted kinetic run values in Figure 5.21.

In the table below, we give the identified rate constants for the desorption kinetics, that is, r_{cs} and r_{qs}. We also include in the table the ratio

$$\frac{r_{qd}}{r_{cd}}$$

which should equal

$$\frac{Solution\ volume}{Clay\ mass} = \frac{20ml}{.02g} = 1000\frac{ml}{g}$$

The deviation from this ratio is an indication of difficulty in maintaining conservation of mass. We will also include the starting mass and the finishing mass predicted by the calibrated model. We used the average of the measurements at each time.

	r_{cd}	r_{qd}	r_{qd}/r_{cd}	Starting Mass (μg)	Ending Mass (μg)
Zinc on a Mississippi clay	1.4555×10^{-5}	.0198	1360.31	31.7102	27.4457

Although we seem to be conserving mass here, there are clearly some problems. We provide a phase plane diagram with the desorption isotherm, the kinetic run data, and the predicted kinetic run values in Figure 5.22.

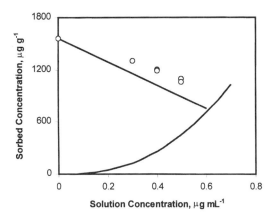

Figure 5.22   Zinc desorption isotherm, theoretical and experimental desorption kinetics on a Mississippi smectite.

MATERIALS AND METHODS

Pretreatment of Smectites

Wyoming and Mississippi smectites were pretreated to remove soluble salts, carbonate, organic matter, and iron oxides and then separated by sieving and centrifugation into sand-, silt-, and clay-sized fractions (Dixon and White, 1977; Jackson, 1956). Iron oxides in the coarse clay were removed with the dithionite-citrate bicarbonate method (Dixon and White, 1977; Jackson, 1956). The clay-sized (< 0.2 mm) particles were separated after a number of resuspensions and centrifugation at 750 rpm for 3.5 minutes in an International model K centrifuge with a #266 centrifuge rotor (Dixon and White, 1977; Jackson, 1956). The clay-sized smectites were saturated with Ca by washing the clay with $0.5\ M$ $CaCl_2$ four times. The free Ca was washed with deionized water until no free Cl was detected.

Clay Properties

The specific surface area (SSA) was determined by EGME (ethylene glycol monoethyl ether) adsorption (Ratner-Zohar et al., 1983). Cation exchange capacity (CEC) was analyzed by procedures for Ca/Mg cation exchange (Dixon and White, 1977).

Adsorption and Desorption

Twenty mg Ca-smectite was shaken with 10 ml of $0.005\ M$ $Ca(NO_3)_2$ solution containing $Zn(NO_3)_2$ from 0.2 to 300 mg Zn L^{-1} for 2 weeks. The suspension was centrifuged. The supernatant was filtered with 0.2 µm microfilter. For the adsorption kinetics study, 20 mg Ca-smectite was shaken with 10 ml of 0.005 M $Ca(NO_3)_2$ solution containing initially 5, 10, 25, and 100 mg Zn L^{-1} for 30 min, 2 h, 8 h, 24 h, 4 d, 1 wk, and 2 wk, respectively. The Zn-smectites, after 2 wk of adsorption with 5, 25, and 100 mg Zn L^{-1} initial Zn concentrations, were shaken with 20 ml deionized water for 10 min, 2 hr, 1 d, 1 wk, and 2 wk. The supernatants were collected for Zn measurements by atomic absorption spectrometer after centrifugation and filtration.

CONCLUSION

We have shown that the use of phase plane analysis from the qualitative theory of differential equations and dynamical systems can be a powerful tool in making judgments about what our experimental data are showing us. In particular, such analysis can be a great help in determining when our models are wrong. We have also seen how a careful use of differential equations can allow us to build new models which are consistent and natural.

REFERENCES

Blanchard, P., R. L. Devaney, and G. R. Hall, 1998. *Differential Equations,* Brooks/Cole, Boston.

Borrelli, R. L. and C. S. Coleman, 1998. *Differential Equations: A Modeling Perspective,* John Wiley & Sons, New York.

Burden, R. L. and J. D. Faires, 1997. *Numerical Analysis,* 6th edition, Brooks/Cole, Pacific Grove.

Coddington, E. A. and N. Levinson, (1955) *Theory of Ordinary Differential Equations,* McGraw-Hill, New York.

Dixon, J. B. and G. W. White, 1977. *Soil Mineralogy Laboratory Manual.* Soil and Crop Sci. Dept., Texas A&M Univ., College Station, TX.

Hirsch, M. W. and S. Smale, 1974. *Differential Equations, Dynamical Systems, and Linear Algebra,* Academic Press, New York.

Jackson, M. L., 1956. *Soil Chemical Analysis — Advanced Course,* published by the author, Dept. of Soils, Univ. of Wisconsin, Madison, WI.

Kingery, W. L., F. X. Hah, K. O. Willeford, and S. F. Oppenheimer. 1998. Sorption and Activity of Extra-Cellular Enzymes on Smectite Clays, p. 12. In *Extended Abstracts of Annual Meetings of the Clay Minerals Society,* Cleveland, OH.

The Math Works Inc., *The Student Edition of MATLABT,* Prentice Hall, New Jersey, 1992.

Oppenheimer, S. F., 1995. A model for batch tests, *Canadian Applied Mathematics Quarterly.* 3, 1, 89-98.

Oppenheimer, S. F., W. L. Kingery, K. Willeford, and F. X. Han. 2001. A quadrature technique for reaction rate identification. *Nonlinear Analysis, Series B: Real World Applications.* 2:135-144.

Ratner-Zohar, Y., A. Banin, and Y. Chen, 1983. Oven drying as a pretreatment for surface area determination of soils and clays, *Soil Soc. Am. J.* 47:1056-1058.

Selim, H. M. and M. C. Amacher, 1997. *Reactivity and Transport of Heavy Metals in Soils,* CRC Press Inc., Boca Raton, FL.

Showalter, R. E. and M. Peszynska, 1998. A transport model with adsorption hysteresis, *Differential and Integral Equations.* 11, 327-340.

CHAPTER 6

Kinetic Study of Trace Metal EDTA-Desorption from Contaminated Soils

A. Bermond and J.P. Ghestem

INTRODUCTION

Some Features of the Chemical Extraction of Soil Trace Metals

Regardless of their origins and the reasons for the increase in their concentration in soils, trace metals are liable to contaminate food chains by migrating toward groundwater or by accumulating in plants. To predict the persistence and potential mobility and bioavailability of these elements, single and/or sequential extractions involving chemical reagents have been widely used. In the first case, the extracted amounts of trace metals by a given chemical reagent are supposed to correlate with the amounts accumulated in plants. In the second case, sequential extractions are mainly supposed to provide the localization of trace metals in soils: the possible mobility and bioavailability are the result of the reactivity of trace metals in soils, which depends on their localization in different soil components, which is now usually called *speciation*. Note that, by also using the term *localization,* speciation of trace elements in soils is not fully defined as it is usually (Ure, 1991) by the different physicochemical forms of the same element.

From a general point of view, chemical extraction methods are more sensitive compared to physical methods of speciation and what we call sequential extraction consist of using successively different chemical reagents for extraction of trace elements from given soil compartments (Belzile et al., 1989; Nirel and Morel, 1990), terminating with their quantification in the extraction phase (normally when equilibrium has been reached) (Whalley and Grant, 1994; Bermond and Malenfant, 1990; Belzile et al., 1989; Nirel and Morel, 1990). Several reagents are generally used, and some classical chemical reagents which involve different chemical reactions (complexation, oxidation, reduction, etc.), are given in Table 6.1.

1-56670-531-2/01/$0.00+$1.50

Table 6.1   Some Chemical Reagents Used in Soil Trace
            Element Speciation Studies

Chemical Reagent	Expected Soil Compartments
NH_2OH $H_2C_2O_4$	Fe, Mn oxides
H_2O_2 NaClO	Organic matter
EDTA $P_2O_7^{4-}$	Oxides, calcareous, organic matter

However, it is generally acceptable that these protocols cannot supply a reliable estimate of the speciation of trace elements in soils, particularly for thermodynamic reasons (measurements made at equilibrium) (Ure, 1991; Whalley and Grant, 1994; Bermond and Malenfant, 1990), so that a sequential extraction procedure is now considered as only operationally defined and gives limited information on associations of trace metals with soil constituents. We will give here a brief overview on the possible artifacts of sequential extractions, focusing on the example of iron which is one of the soil compartments recognized as able to adsorb trace elements.

First, the supposed selectivity of the chemical reagents used in sequential extractions is not really fulfilled. For example, the solubility of iron oxide is given on Figure 6.1. It has been calculated using thermodynamic data and varies when pH changes and/or in the presence of different iron (III) complexants used in sequential extractions, such as pyrophosphate or oxalate acid. This is a well-known result. It can be easily generalized: no chemical reagent, due to its own pH compared to the soil sample's pH or due to its complexing properties, may be assumed to be specific to one soil compartment.

Second, as previously stated, quantification of trace metals in the extraction phase, which terminates each step, is normally performed when equilibrium has been reached; kinetics are not taken into account. As an example, the monitoring of iron during the hydrogen peroxide reaction (Figure 6.2), which in sequential extraction is supposed to give an estimation of trace metals bound to organic matter, indicates undoubtedly that readsorption phenomena are involved during this step. These phenomena are time dependent and, from a general point of view, may not be deduced from equilibrium measurements. This explains why trace metals extracted in a given step of a sequential extraction procedure may be underestimated.

Objectives

According to the considerations above, it appears necessary to consider other methods of determining the associations of trace metals in soils in more detail. As the time dependence of extraction phenomena is not often taken into account when chemical methods of speciation are performed, it could be interesting to consider kinetic aspects of chemical extractions that also may characterize the stability of the various trace metal-soil constituent associations. Moreover, this parameter is probably strongly involved when mobility or bioavailability of trace elements in soils has to be determined; it can be assumed that, in a large number of natural cases,

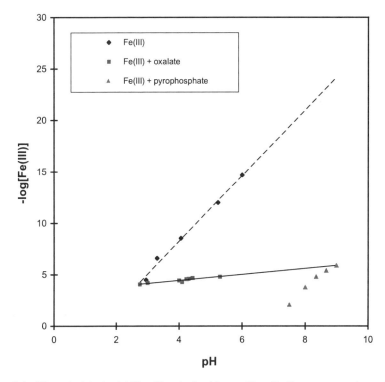

Figure 6.1 The calculated solubility of iron hydroxide vs. pH and in the presence of complexing reagents.

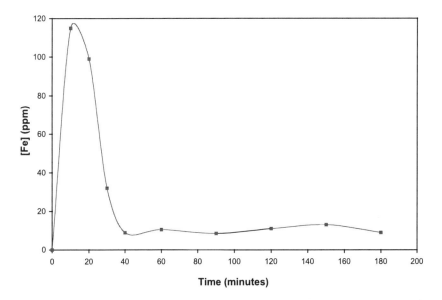

Figure 6.2 The monitoring of extracted iron during the soil organic matter–hydrogen peroxide attack (Château-Thierry sample).

chemical equilibration times are long compared to rates of uptake by organisms. In other words, the extraction of trace metals into the soil solution may be the limiting step for the bioavailability of these elements.

Our aim was first to study the kinetics of extraction of trace metals from polluted soil samples and then to assess the feasibility of using these kinetic data for proposing a fractionation of these trace metals.

KINETIC STUDY OF THE EXTRACTION OF TRACE METALS FROM CONTAMINATED SOILS

Introduction

A number of different kinetic models have been applied to describe the extraction of various ions from soils (Yu and Klarup, 1994). The multiple first-order reaction model has been widely studied by soil scientists. In this model, different first-order reactions are ascribed to discrete types of binding sites, and it is assumed that there are multiple simultaneous first-order reactions and their rates are independent.

From the point of view of the speciation of trace metals, an approach to the speciation using this model or more complex models is recent and has not been extensively applied to soil samples. This approach is generally used with water samples (Chakrabarti et al., 1994; Lu et al., 1994) and sometimes with model compounds such as fulvic acids (Cabaniss, 1990; Lavigne et al., 1987) or humic acids (Rate et al., 1992; Bonifazi et al., 1996), and iron or manganese oxides (Gutzman and Langford, 1993; Backes et al., 1995; Langford and Gutzman, 1992). In this case, most of the authors were more interested in the determination of the global constants of complex dissociation. The multiple first-order reaction model or other various mathematical models and resolution methods were applied depending on the number of compartments used to describe the extraction experiments (Chakrabarti et al., 1994; Lu et al., 1994; Cabaniss, 1990; Lavigne et al., 1987; Rate et al., 1992; Bonifazi et al., 1996; Gutzman and Langford, 1993; Backes et al., 1995).

As previously stated, numerous chemical reagents, particularly complexing reagents (Sposito et al., 1982; Stover et al., 1976; Haynes and Swift, 1983; Borggaard, 1976; Kedziorek and Bourg, 1996) are used in soil science to extract trace metals. According to the second part of our objective of fractionation of soil trace metals, it seemed necessary that the chosen reagent for this study should be able to take into account trace metals bound to most of the chemical soil compartments (organic matter, iron oxides, etc.). For this reason, the chosen reagent was EDTA, because it is well known to extract trace metals from the different soil compartments.

EDTA (ethylenediaminetetraacetic acid) is widely used in soil science for different purposes and has also been reported in the environment (Borggaard, 1976). On the one hand, EDTA is one of the major chemical reagents used to extract trace elements from soil samples in order to predict their bioavailability or mobility (Ure, 1996; Shuman, 1985; Beckett, 1988; Lake et al., 1984). EDTA has also been used

Table 6.2 Physical-Chemical Characteristics of the Studied Soil Samples

	Couhins	Evin	Chat.-Thierry	Mortagne
pH (10/25 ml)	7.1	7.9	7.8	6.3
C. organique (%)	2.2	1.8	0.85	Nd
Argile (%)	2.9	16.5	13.8	Nd
Carbonate (g.kg⁻¹)	—	8	10	Nd
Cd_{tot} ($\mu g.g^{-1}$)	94.9	23	3.5	71.3
Zn_{tot} ($\mu g.g^{-1}$)	151.0	1415	123	10,150
Pb_{tot} ($\mu g.g^{-1}$)	44.8	1120	34	2460
Cu_{tot} ($\mu g.g^{-1}$)	45.3	43.5	33.4	225
Ni_{tot} ($\mu g.g^{-1}$)	245.80	—	—	—
Fe_{tot} ($\mu g.g^{-1}$)	3526.0	20,900	18,142	11,060
Mn_{tot} ($\mu g.g^{-1}$)	69.61	411	587	540
Ca_{tot} ($\mu g.g^{-1}$)	2500	7600	—	3720
Mg_{tot} ($\mu g.g^{-1}$)	212	3300	2880	1320

for many years as a chelating agent for supplying micronutrient cations for plants (Lindsay and Norvell, 1969; Norvell and Lindsay, 1969). Lastly, this species may also be used to remediate contaminated soil. On the other hand, EDTA is one of the most widely used industrial complexing agents (photographic industry, textile and paper manufacturing, industrial cleaning) and persists as a major pollutant in rivers and groundwaters (Novack et al., 1996).

Material and Methods

Soil Samples

For this study, we mainly used four soil samples. Two samples are agricultural soils and have received sewage sludges (Couhins and Chateau-Thierry). The two other samples come from industrial waste land (Evin and Mortagne). Some physicochemical characteristics of these samples are given in Table 6.2. Before use, samples were dried at 20°C and sieved (2 mm).

Choice of Extraction Conditions

In order to determine the extraction conditions for this work, we performed a study of the extraction of trace metals vs. acidity and EDTA concentration (Ghestem and Bermond, 1998). The example of copper EDTA extraction is given in Figure 6.3.

For copper, and for other trace metals such as Cd, Pb, and Zn, it seems that the shape of the extraction curve does not depend very strongly on the soil sample and that the acidity and the concentration of EDTA are in fact the main parameters to be considered. Finally, if we consider the concentrations of the added ligand and the variation of the amount of these extracted trace metals vs. pH, we can observe two cases that correspond or not to an excess of EDTA.

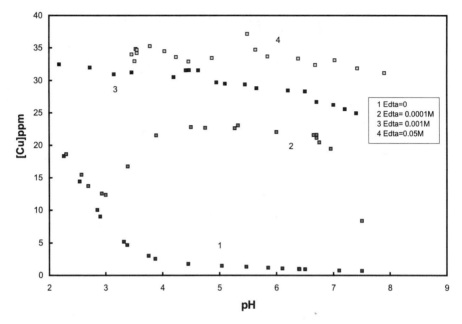

Figure 6.3 The EDTA-desorption of copper vs. pH and different levels of EDTA (Couhins sample).

Lack of EDTA

When EDTA is not in excess (0.0001 M) the extraction of trace elements as a function of pH takes a shape in which it is clearly possible to distinguish three zones from acid to neutral media:

- When pH is at a low level (acid medium) the amount of extracted trace elements decreases when pH decreases, until it joins the values of the blank curve. It could be involved in the adsorption of the complexes MY^{2-}, as described in the literature (Bowers and Huang, 1986; Girvin et al., 1993) onto positive sites of the soil samples. It could also be involved in a competition for these acidity levels between cations and H^+ for the reaction with EDTA (H_2Y^{2-} form). This results in a diminution of the formation of MY^{2-} through the complexation reaction which is progressively replaced by Reaction 6.1.

$$M^{2+}_{soil} + 2H^+ \Leftrightarrow 2H^+_{soil} + M^{2+} \tag{6.1}$$

If we define pHa as the pH corresponding to the beginning of the 2nd zone, it can be noticed that the order of pHa is Cd > Zn > Pb = Cu.
- The second zone ranges from pHa (depending on the considered metal) to approximately neutral pH; in this case, the EDTA extraction appears to be pH independent; the reaction presumed to be involved is Reaction 6.2, which is not pH dependent:

$$M^{2+}_{soil} + H_2Y^{2-} \Leftrightarrow 2H^+_{soil} + MY^{2-} \tag{6.2}$$

- When pH is approximately greater than 6, a decrease in the extracted amount of trace metals is observed; this could be explained by a competition between trace elements and some other cations in the complexation reaction with EDTA, as described in the literature (Nabhan et al., 1977; Lucena et al., 1988), but it seems also to involve chemical Reaction 6.3, which is pH dependent and takes into account the change in the EDTA form or, in other words, the acidity constant of H_2Y_2/HY ($-\log K = 6.4$) (Ringbom, 1967):

$$M^{2+}_{soil} + HY^{2-} + H^+ \Leftrightarrow 2H^+_{soil} + MY^{2-} \tag{6.3}$$

Excess of EDTA

Two of the other studied concentrations of EDTA (0.001 M and 0.05 M) correspond generally to an excess of the complexing agent.

For trace elements (Cu, Pb, Cd and Zn) and for the lower pH values, no effect of EDTA is observed. This means that the curves join the blank curve and that no complexation occurs, due to the low pH and the acid properties of Y^{4-} (Ringbom, 1967). Reaction 6.1 is in this case the predominant reaction. Except for these acidity levels, it can be approximated that the EDTA extraction of trace metals is not pH dependent. These results are consistent with Reaction 6.2, which could be involved in the pH range from 2.5 to 6.

Finally, according to this study the chosen extraction conditions were the following:

- In order to have a large excess of EDTA (pseudo first-order reaction from a kinetic point of view), the initial concentration was 0.05 M and the ratio m/v was equal to 0.02; this excess provides free EDTA and, for this reason, released metals may be considered as complexed.
- The acidity was near neutrality (pH = 6.5) (which permits us to suppose that H^+ ions did not extract trace metals). pH changes were monitored and did not occur due to the buffering capacity of the excess of EDTA.

Experiments

From an experimental point of view, the extraction of trace metals was performed in a glass receptacle with 20 g of sample and 1 L of the EDTA solution (0.05 M) at pH 6.5. These conditions correspond to a ratio m/v equal to 0.02 g/ml. The solution was stirred with a mechanical stirrer at a speed of about 300 rpm.

For kinetics, time zero corresponds to the addition of EDTA to the soil sample. The monitoring of extracted metals vs. time is performed over 24 hours. Twenty-four samples of the solution are taken at the following intervals: 1, 2, 3, 4, 5, 6, 8, 10, 20, 30, 40, 50 min and 1, 2, 3, 4, 5, 6, 7, 8, 9, 24, 24.5, 25 h. Samples were then filtered with Millipore membranes (porosity = 0.45 μm). The pH was measured and the samples were kept at + 4°C prior to trace metal analysis.

All solutions used were made from analytical-grade salts or titrisol solutions and milliQ water. Cations in solution were measured using a VARIAN Spectra 250 Plus atomic absorption spectrometer equipped with an air-acetylene flame and external standards. Analytical performances (sensitivity, detection thresholds) in the EDTA

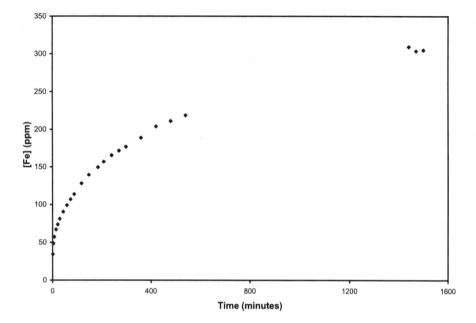

Figure 6.4 The monitoring of EDTA-extracted iron vs. time (Evin sample).

solutions are not different from those obtained in simple matrices. Furthermore, we used the standard addition method to verify that there was no interference, and in all cases we made extraction blanks. Under these conditions, the coefficients of variation obtained on extraction results with the reagents used — including the variability of the soil sample — are satisfactory and range between 2 and 6%, depending on the metal (for four repetitions).

Kinetic Data

EDTA is a strong extractant due to its complexing properties and may dissolve some soil compounds as iron hydroxides or calcareous. As an example, Figure 6.4 shows the monitoring of solubilized iron during EDTA extraction. It can be seen that about 10% of total iron is dissolved within 1 day, but equilibrium would probably require more than a week to be reached.

From the point of view of trace elements, the EDTA extraction curves of zinc and cadmium vs. time are presented on Figures 6.5 and 6.6, respectively. These two examples may be considered as representative of the different kinetics encountered in this study: the first one corresponds to one cation very quickly extracted, while the second one corresponds to a cation more slowly EDTA extracted.

However, from a general point of view, it can be said that the EDTA extraction of trace elements is fast enough, at least for the trace metals studied in this work. Table 6.3 shows that, over a 24-h time span, more than 70% of EDTA-extracted trace metals are extracted in the first 30 min. This suggests that this fraction of trace metals was loosely held by soil particles (Yu et al., 1994).

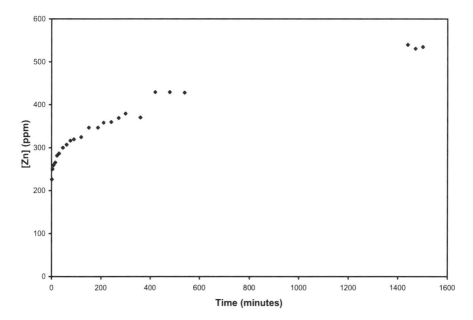

Figure 6.5 The monitoring of EDTA-extracted cadmium vs. time (Evin sample).

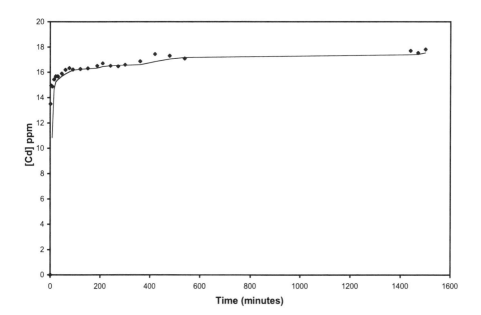

Figure 6.6 The monitoring of EDTA-extracted zinc vs. time (Evin sample).

Table 6.3   The Extraction Ratio $[M]_{0.5}/[M]_{24}$ for Some Trace Metals (four repetitions)

	Cd	Cu	Zn	Pb	Ni
Couhins sample	0.89	0.8	0.8	—	0.73
Evin sample	0.91	0.79	0.60	0.91	—

Principles and Feasibility of the Kinetic Fractionation Method

From the previous kinetic experiments, we tried to study the feasibility of performing an operational kinetic fractionation method applied to soil samples and not truly related to speciation, as it is classically defined. This method should be able to provide, from the monitoring of the extraction of soil trace elements vs. time with a given reagent, the concentrations of the extracted trace metals from only two compartments. This means that we assume that there exist only two sorts of metallic cations, i.e., those quickly extracted (called *labile*) and those less quickly extracted (called *slowly* or *moderately labile*). This assumption for the proposed fractionation method may be represented by the following reactions with the extracting reagent R (Backes et al., 1995; Langford and Gutzman, 1992):

$$S_1M + R \rightarrow S_1 + MR \tag{6.4}$$

$$S_2M + R \rightarrow S_2 + MR \tag{6.5}$$

where S_1 and S_2 represent, respectively, the labile and slowly labile compartments. We have to keep in mind that these two compartments are only kinetically defined and that they are not at all related to the classical soil compartments, such as organic matter or iron oxides for example. Note that, if we consider the number of compartments used in the literature for the application of kinetics even to model compounds, the choice of only two compartments, as we did for a soil sample, seems to be a challenge; it is, however necessary in order to get useful results. More than two compartments would be difficult to use.

Because the reagent EDTA used in this study is in excess, we can consider that the reactions are pseudo first-order reactions. Thus, according to the multiple first-order reaction model, the rate of desorption of an ion from the soil particles is given by Equation 6.6:

$$C(t) = \Sigma C_i\left(1 - e^{-k_i t}\right) \tag{6.6}$$

where
   C_i denotes the ion concentration bound to site i;
   k_i denotes the kinetic constant related to the ion extraction from site i;
   $C(t)$ is the concentration of the extracted ion (measured in the solution) at time t.

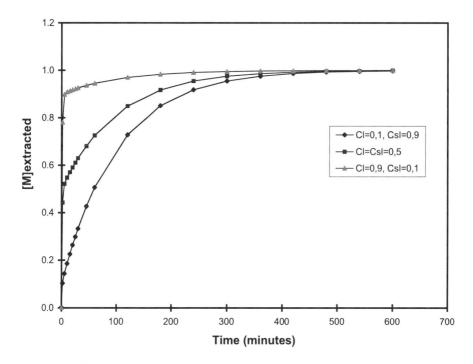

Figure 6.7 Simulation of trace metal extraction according to the equation ($K_l = 1$, $K_{sl} = 0.01$).

If we apply a two first-order reaction model, according to our hypothesis (only a labile and a slowly labile compartment from Reactions 6.4 and 6.5), Equation 6.6 simplifies into Equation 6.7:

$$C_{tot} - C(t) = C_L e^{-k_L t} + C_{SL} e^{-k_{SL} t} \qquad (6.7)$$

where

C_{tot} denotes the total concentration of EDTA-extracted cation;

C_L and C_{SL} denote, respectively, the concentration of the labile cation and the concentration of the slowly labile cation.

A simulation with arbitrary data of the evolution of the concentration C(t) of the extracted metal, using Equation 6.6, is given in Figure 6.7. In this example, where the ratio of labile to slowly labile cation concentration was varied, it can be seen that this ratio determines the shape of the extraction curve: the greater the slowly labile concentration, the slower the extraction of the metallic cation. It is obvious that a similar conclusion would be drawn from the simulation of the influence of the ratio of the two kinetic constants, k_L and k_{NL}. These simulated curves exhibit a shape which is not different from the shape of the experimental extraction curves.

The feasibility of this proposed fractionation method was studied on the four soil samples (Couhins, Evin, Mortagne, and Evin). In order to determine the two

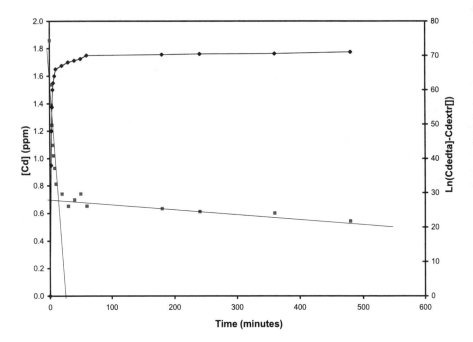

Figure 6.8   The EDTA-extraction of zinc and the logarithmic transformation using Equation 6.9 (Evin sample).

concentrations, C_L and C_{SL}, we used a graphic transformation of the extraction curve. For the longer extraction time, we may ignore the term corresponding to the labile fraction of cations, which has already been released. In fact, it is simply supposed that the two reactions involved in the model are, from the kinetic point of view, different enough, which finally gives Equation 6.8:

$$C_{tot} - C(t) \approx C_{SL}\, e^{-k_{SL}t} \tag{6.8}$$

or in a logarithmic form:

$$Ln\big(C_{tot} - C(t)\big) = Ln\big(C_{SL}\big) - k_{SL}t \tag{6.9}$$

If these previous assumptions are correct, the extraction curves for longer times may be transformed into one straight line. Two examples of the application of this approach to the zinc and cadmium extraction curves from soil samples are given in Figures 6.8 and 6.9. It can be seen that a good agreement with the previous assumptions is obtained, which allows us to calculate from the straight line (corresponding to the longer times) the concentration C_{SL}, from which we can deduce C_L.

Finally, these two examples indicate that the proposed model for the operational speciation of trace metals with only two compartments may be used and gives

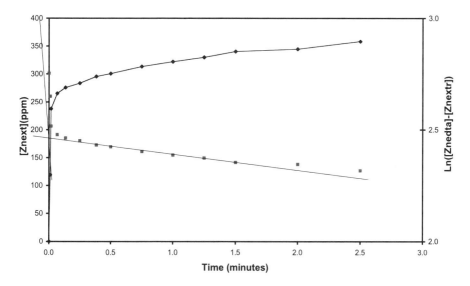

Figure 6.9 The EDTA-extraction of cadmium and the logarithmic transformation using Equation 6.9 (Château-Thierry sample).

Table 6.4 Repeatability of the Labile and Slowly Labile Trace Metal Concentration Determination (Evin sample)

	Cd		Zn		Cu		Pb	
C_l and C_{sl}	66	10	19	20	41	15	78	12
Stand. dev.(%)	0.2	11	3	2	1	5	3	5

Table 6.5 Repeatability of the Labile and Slowly Labile Trace Metal Concentration Determination (Couhins sample)

	Cd		Zn		Cu		Ni	
C_l and C_{sl}	54	22	41	17	41	22	44	31
Stand. dev.(%)	6	7	5	8	4	3	8	1

satisfactory results: the experimental data obtained during the monitoring of EDTA-extracted metals from polluted soil samples may therefore be described, from the kinetic point of view, with only two compartments. Similar results were obtained for other cations and samples.

In order to study the quality of the kinetic speciation method, its repeatability was then determined. The results obtained for two soil samples are presented in Tables 6.4 and 6.5. It can be seen that the repeatability of the speciation method ($C_L\%$ and $C_{SL}\%$, i.e., amounts of trace metals respectively called labile and slowly labile) is less than 10% whatever the cation or the soil sample is. This seems to be a very good result if we consider that soil samples are more or less heterogeneous.

Table 6.6   Range of the Labile and Slowly Labile Trace Metal Concentration Expressed as a Percentage of the Total Concentration (all samples)

	Cd	Cu	Zn	Pb[a]	Ni[b]
$[M_l]$ min	52	64	20	58	22
$[M_l]$ max	68	28	40	78	—
$[M_{sl}]$ min	10	16	18	14	16
$[M_{sl}]$ max	22	30	38	24	—

[a] 3 determinations
[b] 1 determination

We will consider now the concentrations of C_L and C_{SL} (Table 6.6) calculated from the experiments with at least two replicates and dealing directly with the fractionation determination for the studied trace metals and soil samples.

We can first recall that for these samples, EDTA behaves as a strong extraction reagent; for most of the metals, it extracts more than 50% of the total amount. It can also be seen from these results that the amount of labile cations is always significant (with the exception of zinc): C_L often represents more than 40% of the total amount and may reach 80% (Pb from the Evin sample). As a consequence, the slowly labile fraction of the EDTA-extracted metals is weak; it may be about 10% of total cadmium, for example (Evin sample).

If we examine each metal, some tendencies can be observed. Lead and cadmium are found to a great extent in a labile form (more than 50% of the total). This is a classical result for cadmium, which is generally said to be mobile, but not for lead. According to these results, the copper labile fraction ranges from 28 to 64%, depending on the soil sample, while nickel and zinc seem to be the less labile cations.

Finally, one question may arise: does a relationship exist between kinetic data and thermodynamic data? To answer this question, we studied the correlation between the calculated labile trace metal fraction and the EDTA-extractable amount (Figure 6.10). The correlation is poor ($r^2 = 0.6$), which means that kinetic data may also provide different and useful information in predicting reactivity of trace metals in soils.

CONCLUSION

The aim of this study was to assess the feasibility of a soil trace metal fractionation using a kinetic approach. It has been shown with different soil samples that the monitoring of EDTA-extracted trace metals vs. time may be used to differentiate fast-extracted cations — called *labile* — and slowly extracted cations — called *slowly labile*; this approach requires some simple kinetic hypotheses to establish the equation of the extraction curves and the calculation of the labile and slowly labile fraction. This fractionation method does not involve classical soil compartments such as organic matter or iron oxides, for example, but is mainly an operationally defined method.

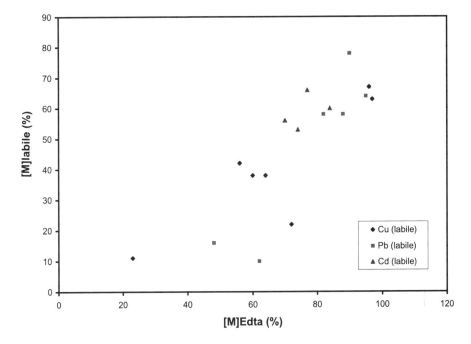

Figure 6.10 Correlation of the calculated labile trace metal fraction vs. the EDTA-extractable amount (all samples).

According to these preliminary results, which show the feasibility of a kinetic approach to the fractionation of soil trace metals, this method has now to be applied to other samples in order to assess its possible use in terms of prediction of the behavior of trace elements. For example, the correlations between the labile fraction and the bioavailability of the trace elements are to be studied.

REFERENCES

Backes C.A., McLaren R.G., Rate A.W., and Swift R.S., 1995. Kinetics of Cd and Co desorption from iron and manganese oxides, *Soil Sci. Soc. Am. J.* 59, 778-785.

Beckett P.H.T., 1988. The use of extractants in studies on trace metals in soils sewage sludges, and sludge-treated soils. In *Advances in Soil Science,* Springer Verlag, New York, 144-171.

Belzile N., Lecomte P., and Tessier A., 1989. Testing readsorption of trace elements during partial chemical extractions of bottom sediments, *Environ. Sci. Technol.* 23(8), 1015-1020.

Bermond A. and Malenfant C., 1990. Estimation des cations métalliques liés à la matière organique à l'aide de réactifs chimiques: approche cinétique, *Science du Sol.* 28(1), 43-51.

Bonifazi M., Pant B.C., and Langford C.H., 1996. Kinetic study of the speciation of Cu(II) bound to humic acid, *Environ. Technol.* 17, 885-890.

Borggaard O.K., 1976. The use of EDTA in soil analysis, *Acta Agriculturae Scandinavica.* 26, 144-150.

Bowers A.R. and Huang C.P., 1986. Adsorption characteristics of metal-EDTA complexes onto hydrous oxides. *J. Colloid and Interface Sci.* 110, 575-591.

Cabaniss S.E., 1990. pH and ionic strength effects on Ni-fulvic acid dissociation kinetics, *Environ. Sci. Technol.* 24(4), 583-588.

Chakrabarti C.L., Lu Y., Gregoire D.C., and Back M.H., 1994. Kinetic studies of metal speciation using chelex cation exchange resin: application to Cd, Cu, and Pb speciation in river water and snow, *Environ. Sci. Technol.* 28, 1957-1967.

Ghestem J.P. and Bermond A., 1998. EDTA extractability of trace metals in polluted soils: a chemical-physical study. *Environ. Technology.* 19, 409-416.

Girvin D.C., Gassman, P.L., and Bolton H., 1993. Adsorption of aqueous cobalt ethylendi-aminetetraacetate by Al_2O_3. *Soil Sci. Soc. Am.* 57, 47-56.

Gutzman D.W. and Langford C.H., 1993. Kinetic study of the speciation of Cu(II) bound to a hydrous ferric oxide, *Environ. Sci. Technol.* 27(7), 1388-1393.

Haynes R.J. and Swift R.S., 1983. An evaluation of the use of EDTA and DTPA as extractants for micronutrients in moderately acid soils, *Plant & Soil.* 74, 111-122.

Kedziorek A.M. and Bourg A.C.M., 1996. Acidification and solubilisation of heavy metals from single and dual-component model system. *Appl. Geochem.*, 11, 299-304.

Lake D.L., Kirk P.W., and Lester J.N., 1984. Fractionation, characterisation, and speciation of heavy metals in sewage sludge and sludge-amended soils: a review. *J. Environ. Technol. Qual.* 13, 175-183.

Langford C.H. and Gutzman D.W., 1992. Kinetic studies of metal ion speciation, *Analytica Chimica Acta.* 256, 183-201.

Lavigne J.A., Langford C.H., and Mak M.K.S., 1987. Kinetic study of speciation of Ni(II) bound to a fulvic acid, *Analytical Chemistry.* 59, 2616-2620.

Lindsay W.L. and Norvell W.A., 1969. Equilibrium relationships of Zn^{2+}, Fe^{3+}, Ca^{2+} and H^+ with EDTA and DTPA in soils. *Soil Sci. Soc. Am. Proc.* 33, 86-91.

Lu Y., Chakrabarti C.L., Back M.H., Gregoire D.C., and Schroeder W.H., 1994. Kinetic studies of aluminium and zinc speciation in river water and snow, *Analytica Chimica Acta.* 293, 95-108.

Lucena J.J., Garate A. and Carpena O., 1988. Theoretical and practical studies on chelate-Ca-pH system in solution. *J. Plant Nutrition.* 11, 1051-1061.

Nabhan H.M., Vandereelen J., and Cottenie A., 1977. Chelate behaviour in saline alkaline soil conditions. *Plant & Soil.* 46, 603-618.

Nirel P.M.V. and Morel F.M.M., 1990. Pitfalls of sequential extractions, *Water Research.* 24(8), 1055-1056.

Norvell W.A. and Lindsay W.L., 1969. Reactions of EDTA complexes of Fe, Zn, Mn and Cu with soils. *Soil Sci. Soc. Am. Proc.* 33, 86-91.

Novack B., Lützenkirchen J., Behra P., and Sigg L., 1996. Modeling the adsorption of metal-EDTA complexes onto oxides. *Environ. Sci. Technol.* 30, 2397-2405.

Rate A.W., McLaren R.G., and Swift R.S., 1992. Evaluation of a log-normal distribution first order kinetic model for Cu(II)-Humic acid complex dissociation, *Environ. Sci. Technol.* 26, 2477-2483.

Ringbom, A., 1967. *Les complexes en chimie analytique.* Dunod Ed. Paris (1967).

Shuman L.M., 1985. Fractionation method for soil micro elements. *Soil Sci.* 140, 11-22.

Sposito G., Lund L.J., and Chang A.C., 1982. Trace metal chemistry in arid zone field soils amended with sewage sludge: fractionation of Ni, Cu, Zn, Cd, Pb in solid phases, *Soil Sci. Soc. Am. J.* 46, 260-264.

Stover R.C., Sommers L.E. and Silviera D.J., 1976. Evaluation of metals in wastewater sludge, *Journal Water Pollution Control Federation.* 48(9), 2165-2175.

Ure A.M., 1991. Trace elements speciation in soils, soil extracts and solutions, *Mikrochimica Acta*. 2, 49-57.

Ure A.M., 1996. Single extraction schemes for soil analysis and related applications. *Sci. Total Envir.* 178, 3-10.

Whalley C. and Grant A., 1994. Assessment of the phase selectivity of the European Community Bureau of Reference (BCR) sequential extraction procedure for metals in sediment, *Analytica Chimica Acta*. 291, 287-295.

Yu J. and Klarup D., 1994. Extraction kinetics of copper, zinc, iron and manganese from contaminated sediment using disodium ethylenediaminetetraacetate. *Water Air Soil Pollut.* 75, 205-235.

CHAPTER 7

Soil Properties Controlling Metal Partitioning

Christopher A. Impellitteri, Herbert E. Allen, Yujun Yin, Sun-Jae You, and Jennifer K. Saxe

INTRODUCTION

Establishment of soil screening levels for risk assessment for both bioavailability and the protection of groundwater relies on an understanding of the lability of chemicals in soils. It has been well documented that the lability (mobility and bioavailability) of heavy metals varies significantly with soil properties for a similar total soil metal concentration. Thus, identification of the major soil parameters affecting metal lability in soils is requisite to predication of metal behavior and establishment of appropriate soil screening levels.

The partitioning coefficient (also known as the distribution coefficient) in soils is a convenient and effective way of comparing the behavior of various contaminants in different soils. The partitioning coefficient (K_d) for a metal in a soil is the concentration of the metal associated with the soil solid divided by the concentration of the metal in soil solution. The availability of metals to organisms, and therefore the toxicity of metals to organisms, is more closely related to partitionable metal rather than total metal concentrations in soils. A soil with high total metal concentrations may be relatively harmless to soil organisms if conditions are such that the desorption/dissolution of metals from soil solids is restricted. Conversely, soils with lower total metal concentrations may affect soil organisms to a great extent if soil conditions are optimal for metal dissolution and desorption.

In this chapter, we will present a review of past and current research concerning metal partitioning in soils, discussion on important parameters affecting partitioning of metals in soils, and a case study on the natural and methodological factors that can affect results in metal partitioning studies.

THE PARTITIONING COEFFICIENT: DEFINITION

The partitioning coefficient has been widely used in modeling the fate and transport of metals (and also other inorganic and organic contaminants) in the environment because of the ease of measurement, and of the large amount of data concerning total concentrations. It is assumed that there is an equilibration of total metal in the solution ($M_{solution}$) and total metal in the soil solid phases (M_{soil}):

$$M_{solution} \leftrightarrow M_{soil} \tag{7.1}$$

The partitioning coefficient relates the concentration of metal in the two phases:

$$K_d = \frac{[M_{soil}]}{[M_{solution}]} \tag{7.2}$$

The metal can be bound to a number of components of the soil, including particulate organic matter and iron and manganese oxides. In the solution phase, the metal can exist as the free metal ion and as inorganic and organic complexes. Thus, we can express the partitioning coefficient

$$K_d = \frac{[M_{soil}]}{[M_{solution}]} = \frac{[M-POM]+[M-FeOx]+[M-MnOx]+...}{[M^{2+}]+[MOH^+]+[MCO_3^0]+[M-DOM]+...} \tag{7.3}$$

For the partitioning coefficient to be the same for a number of soils requires that the distribution of metal species in the solid phase remain constant and that the distribution of metal species in the solution phase remain constant. Because these distributions vary among soils, the partitioning coefficients likewise vary.

To be better able to predict the partitioning of a metal between the solid and solution requires that the pertinent chemical reactions be explicitly considered. For example, the equilibrium between metal in one solid phase component, such as particulate organic matter, and the free metal ion can be expressed

$$M^{2+} + POM \leftrightarrow M-POM \tag{7.4}$$

for which the equilibrium constant is

$$K_{M-POM} = \frac{[M-POM]}{[M^{2+}][POM]} \tag{7.5}$$

Equation 7.5 will be the same for different soils if the nature of the organic matter is constant. However, methods are not usually available for the measurement of the specific chemical quantities, M^{2+}, M-POM, and POM. If a single chemical form is dominant in the solid phase, then the others can be neglected. In this case, we have concentrated on the organic matter present in the soil as being the component principally responsible for metal partitioning. Similarly, in the solution phase, it is possible to consider principal forms of the metals.

It is also necessary to account for other factors that will be important in controlling the partitioning of metals. The surface binding sites in the particulate organic matter and the metal oxides will react with protons and with other metal ions, in addition to reacting with the metal under consideration:

$$H^+ + POM \leftrightarrow H - POM \qquad (7.6)$$

$$Ca^{2+} + POM \leftrightarrow Ca - POM \qquad (7.7)$$

Likewise, similar reactions in the aqueous phase compete with that for the metal of interest:

$$H^+ + DOM \leftrightarrow H - DOM \qquad (7.8)$$

$$Ca^{2+} + DOM \leftrightarrow Ca - DOM \qquad (7.9)$$

Finally, there is partitioning of the soil organic matter between the solution and solid phases:

$$DOM \leftrightarrow POM \qquad (7.10)$$

Operationally, the basic components of K_d (values of $[M]_{soil}$ and $[M]_{solution}$) can be measured in any number of ways. For example, $[M]_{soil}$ may be measured by nitric acid digestion for "total recoverable" metals (USEPA, 1997), or $[M]_{soil}$ may be measured using a more rigorous HF or $HClO_4$ digestion (USEPA, 1995). $[M]_{solution}$ may be measured by ion specific electrode in salt solution soil extracts (ISE) (Sauvé et al., 1997), by anodic stripping voltammetry in salt solution extracts of soil (Gerritse and Driel, 1984), or by ICP-AES analysis of water extracts (Yin et al., 2000). Thus, it is very important to understand how the K_d value for a particular experiment was constructed before making comparisons between studies. For example, all else being equal, the K_d value for a particular soil will tend to be greater with more rigorous methods of estimating $[M]_{soil}$.

PAST AND CURRENT RESEARCH

Gerritse and van Driel (1984) determined "distribution constants" for Cd, Cu, Pb, and Zn in 33 European temperate soils. The authors define a distribution constant (D) as $\Delta C_s / \Delta C_m$, where ΔC_s = increase in concentration of metal in soil (mg/kg) and ΔC_m = increase in metal in soil extract during equilibration of metal solutions with soils (ranging from 0 to 25 µg metal per gram soil). They discovered significant log-log relationships between distribution constants related to soil organic matter concentration and hydrogen ion content in the soil extracts. Results illustrated that "exchangeable" forms of Pb ranged from 1 to 5%, while similar forms of Cd, Zn, and Cu ranged from 10 to 50%.

Anderson and Christensen (1988) examined 38 Denmark soils and calculated distribution coefficients (K_d) for Cd, Co, Ni, and Zn. They also analyzed the soils for metal oxides, organic carbon, cation exchange capacity, and clay content. Emphasis was placed on the testing of soils with low metal concentrations. Most studies prior to this examined soils that contained relatively high metal concentrations. The soils were equilibrated (24 hours) with metal spiked solutions of $CaCl_2$ (10^{-3} M) and K_d values calculated. They found that pH was the single most important factor governing partitioning of the metals in the study. Clay content and hydrous Fe and Mn oxides were also significant factors. They postulated that soil organic matter might play a role in Cd and Ni removal from solution in these batch adsorption experiments. Last, they proposed that reasonable estimates of the distribution coefficients for the metals in this system may be calculated based on pH alone using empirical regression models.

Jopony and Young (1994) studied equilibrium desorption (14 days) of Cd and Pb in an equimolar (0.005 M) solution of $CaCl_2$ and $Ca(NO_3)_2$. They illustrated the influence of filter pore size on measurement of $[M]_{solution}$. Higher removal of colloidal material from solution resulted in lower apparent $[M]_{solution}$. This would cause K_d values to increase. The authors concluded that K_d (based on total metal divided by free metal ion as calculated by a speciation model) is uniquely pH dependent. They developed equations to predict free Cd^{2+} and Pb^{2+} based on total metal concentration in the soil and soil pH. For the Pb study, 70 soils with varying contamination levels from mine spoils were utilized. For the Cd study, they used a combination of mine spoil polluted soils and sewage sludge amended soils. The study also included uncontaminated soils that were amended with mine spoils.

Effects of the type of extraction used to estimate $[M]_{soil}$ on K_d values were examined in a study by Gooddy et al. (1995). The researchers employed 0.01 M $CaCl_2$, 0.1 M $Ba(NO_3)_2$, and 0.43 M HNO_3 extractions to represent $[M]_{soil}$. K_d values for samples from two soil profiles were calculated for 48 elements using pore water from centrifuged soil samples. $Ba(NO_3)_2$ extracted more metals than $CaCl_2$, yielding higher K_d values. The nitric acid extraction resulted in the highest concentration of metals and gave the highest K_d values. Cd tended to be the most strongly bound metal. The order of decreasing K_d values for elements changed with different extractions. The authors attributed the lack of correspondence between the results and the traditional sequence of binding affinities partly to the high levels of DOC in the soil solutions. They stated that the DOC tends to reduce the sorption of strongly bound ions at small concentrations. The authors also postulate that the partition coefficient will be insensitive to pH change and metal-ion activity if dissolved OM and particulate OM dominate metal binding in solution and to the soil solid phase. This postulation assumes that metal binding between solid and dissolved OM is functionally similar.

Lee et al. (1996) examined the partitioning of Cd on 15 New Jersey soils and found that the partitioning of Cd in these soils was highly pH dependent. The 15 soils were equilibrated (24 h) with 1×10^{-4} M $Cd(NO_3)_2$ at pH values ranging from

3 to 10. The results compared favorably with data from the study by Anderson and Christensen. The relationship underwent further improvement with the inclusion of soil organic matter. By normalizing the K_d with SOM (resulting in the SOM normalized partition coefficient, K_{om}) the correlation coefficient improved from 0.799 to 0.927. The researchers concluded that the diffusion of Cd through organic matter coatings onto underlying sorptive materials was insignificant.

Janssen et al. (1997) examined 20 Dutch soils and used regression analyses to formulate equations relating partitioning of metals (As, Cd, Cr, Cu, Ni, Pb, and Zn) with soil parameters. In this study, partitioning coefficients were constructed based on $[M]_{soil}/[M]_{solution}$ where $[M]_{solution}$ was based on metals in soil pore water extracted by a centrifugation procedure. They concluded that the most influential factor for distribution of Cd, Cr, Pb, and Zn between soil solid phase and pore water is pH. Fe content of the soil most significantly influenced the distribution of As and Cu in these soils. Dissolved organic carbon was the most important factor governing the distribution of Ni. The regression models constructed were verified by analysis of a set of British soils. The predictions of the distribution of metals in the British soils were of lesser quality. This reduction in predictive capability was attributed to the fact that the British soils were more acidic than the Dutch soils used to construct the models.

Sauvé et al. (2000) reviewed studies of metal partitioning and reported that there is large variability in reported soil-liquid partitioning coefficients (K_d) for the metals cadmium, copper, lead, nickel, and zinc. They used multiple linear regression analysis and found that K_d values were best predicted using empirical linear regressions with pH alone or pH and either the log of soil organic matter (SOM) or the log of total soil metal. The importance of both pH and organic matter in controlling the partitioning of metals in soils has been the focus of several studies in this laboratory (Lee et al., 1996; Yin et al., 2000). Future research concerning metal partitioning should include prediction or estimation of the partitionable metal that ultimately may become available to soil organisms. Research in this laboratory currently focuses on potentially plant-available metals by constructing partitioning coefficients using equilibrium-based extractions that most closely relate to plant tissue concentrations. By combining partitioning studies with plant uptake trials, we hope to elucidate information concerning the most important parameters affecting partitionable metals that may become plant available.

FACTORS AFFECTING METAL PARTITIONING IN SOILS

When reporting K_d values for soils, it is of paramount importance that the definitions of $[M]_{soil}$ and $[M]_{solution}$ are given. It is also essential for researchers to identify what forms of metals they wish to describe as being partitionable in a particular experiment. For instance, it may be of more importance to studies focusing on metals in groundwater to include all potentially soluble species of metals on soils. For research focusing on metals that are potentially available to plants, it will be necessary to define a value for $[M]_{solution}$ that most closely relates to forms of metals that can potentially become plant available. Speciation of metals after desorption/dissolution is of critical importance when studying uptake of metals by soil organisms.

Relationships between solution speciation and organism uptake is currently a matter of great debate but is beyond the scope of this chapter. For estimating potentially bioavailable metals, researchers have utilized total metals in water extractions to represent partitionable metals (Janssen et al., 1997), free Cu^{2+} ion as measured by ion-specific electrode in dilute salt extractions (Sauvé et al., 1997), anodic stripping voltammetry labile metals in dilute salt extractions (Sauvé et al., 1998), and free ion concentrations given by speciation programs (Jopony and Young, 1994).

This section will examine factors that affect metal partitioning in soil and also laboratory procedures that affect K_d values. The case study presented will illustrate the effects of both an important soil parameter (OM) and an important laboratory procedure (soil:solution ratio) on K_d values.

The most important variables affecting metal partitioning in soils in nature are the same factors that affect desorption/dissolution of metals in soils. Metals on soil solids may enter the soil solution by desorption and/or dissolution (Evans, 1989; McBride, 1994; Sparks, 1995). Metal precipitates, which may be present at higher concentrations of metal in soil, will dissolve to maintain equilibrium concentrations in the solution phase. Desorption processes primarily depend on the characteristics of the solid, complexation of the desorbing metal, system pH, the ionic strength of solution, the type and species of possible exchanging ions in solution, and kinetic effects (i.e., residence time).

pH

Soil pH is considered the master variable concerning metal behavior in soil systems (McBride, 1994) and is the most important factor affecting metal speciation in soils (Sposito et al., 1982). Generally, desorption of metals is increased as pH decreases. Thus, metals tend to be more soluble in more acidic environments. Solubility of metals may increase at higher pH due to binding with dissolved organic matter (DOM) (Allen and Yin, 1996). The solubility of SOM increases with pH increase (You et al., 1999).

Soil solids with pH-dependent charge tend to deprotonate with increasing pH. Metals in solution can then react at these negatively charged, deprotonated sites. There is also less proton competition for fixed charge sites at higher pH values. Both of these factors contribute to increasing desorption of metals with decreasing pH. These effects of pH are well documented (Farrah and Pickering, 1976; Harter, 1983; Barrow, 1986; Hogg et al., 1993; Temminghoff et al., 1994). At high pH, metals may simply precipitate out of solution onto soil solids (Barrow, 1986).

Ionic Strength

Increased ionic strength in solution generally decreases sorption of cations in soil systems, assuming that surfaces are negatively charged. This results in an inverse relationship between ionic strength and K_d. Egozy (1980) found that Co distribution coefficients decreased as soil solution salt concentration increased. Theoretically, as ionic strength increases, the reactive layer for cation sorption decreases in thickness. Di Toro et al. (1986) found the same results, but the ionic strength effects were overshadowed by high particle concentrations in the batch extractions of quartz and montmorillonite.

Kinetic Effects

Rarely do rates of adsorption equal rates of desorption for metals on soil surfaces. Usually, rates of desorption are much slower than rates of adsorption. This phenomenon could result from a number of processes. The sorption of a cation to a surface may be thermodynamically favorable. The reverse reaction would theoretically require a high activation energy (E_a) to occur and may not be thermodynamically favorable (McBride, 1994). Sorbed metals may undergo rearrangement on the sorbing surface. Backes et al. (1995) suggested that desorption of Cd and Co from Fe-oxides slowed with time due to the movement of the sorbed metals to sites exhibiting slower desorption reactions.

The adsorbed metal may be incorporated into recrystallized structures on the solid surface. Ainsworth et al. (1994) attributed the lack of apparent reversibility of Co and Cd partitioning (hysteresis) to incorporation of these metals into recrystallized Fe-oxide structures. The hysteresis is greater if the sorbing metals are allowed to react longer with the Fe-oxides, resulting in a residence time effect.

Nature of Exchanging Cations

Generally, ions with smaller hydrated radius and/or greater charge will exchange for cations with greater hydrated radius and lesser charge on a surface. This ideal behavior may not be exhibited in situations where there are sites that sterically prefer cations of one size. An example of this preference is given by K^+ ions, which fit snugly in the interlayers of vermiculite. This behavior may not be exhibited where there is a high degree of specificity for a certain ion, such as the specificity of OM for Cu. When studying partitioning of metals by batch extraction using neutral salt solutions (e.g., $CaCl_2$) the effects of the exchanging cation must be considered. If a large number of binding sites are specific for Ca^{2+}, weakly bound metals may be exchanged and K_d values decreased.

Soil Solid Characteristics

Primary minerals (e.g., quartz, feldspar), secondary minerals (e.g., clay minerals), metal oxides (which may be primary or secondary minerals), and organic matter (e.g., detritus) compromise the majority of soil solids. Desorption of metals from clay minerals may be governed by system pH for minerals with predominantly pH-dependent charge, such as kaolinite. System pH will be less important for clay minerals such as montmorillonite where isomorphic substitution gives a permanent negative charge to the mineral (Sparks, 1995). The location of the sorbed metal on or in the clay mineral also plays a role in desorption. If the metal is bound in a collapsed section of a layered phyllosilicate (e.g., vermiculite), desorption occurs more slowly than for the same metal bonded at the surface (Scheidegger et al., 1996). Backes et al. (1995) found that Cd and Co desorption occurred much more readily on Fe-oxides compared with Mn-oxides. Metals incorporated into the structure of recrystallized oxides may reduce the desorption of metal from solid to solution (Ainsworth et al., 1994; Ford et al., 1997). Soil organic matter (SOM) can sorb/chelate metals. McBride (1994) proposed the following order for the chelation of metal by SOM based on Pauling electronegativities:

$$Cu^{2+} > Ni^{2+} > Pb^{2+} > Co^{2+} > Ca^{2+} > Zn^{2+} > Mn^{2+} > Mg^{2+}$$

Metals sequestered in the structure of organic molecules may not readily desorb. The amount of desorption of Pb^{2+}, Cu^{2+}, Cd^{2+}, Zn^{2+}, and Ca^{2+} from peat was much less than the amount sorbed (Bunzl et al., 1976). McBride et al. (1997) noted that 0.01 M $CaCl_2$ extractions contained lower concentrations of Cu than did H_2O extractions on the same soil. This phenomenon was attributed to the diminished solubility of Cu-organic complexes in the presence of Ca^{2+}. Desorption of metals from organic matter is pH dependent as the main functional groups (carboxylic and phenolic) on SOM exhibit pH-dependent charge (Sparks, 1995). SOM may have a greater impact on soils with low inorganic cation exchange capacity (CEC). Elliot et al. (1986) found that removal of SOM from soils reduced sorption of Pb, Cu, Cd, and Zn, but only sorption of Cu and Cd were reduced upon removal of SOM from a soil with high inorganic CEC.

Complexation of Desorbing Metal

Recent work using spectroscopic and microscopic techniques provides a wealth of information concerning relationships between metals and soil solids. Much of the work with extended X-ray absorption fine structure spectroscopy (EXAFS) and X-ray absorption near edge spectroscopy (XANES) reveals specific binding mechanisms and/or evidence of precipitation by metals onto pure solids. For example, Scheidegger et al. (1996), found evidence of Ni bonding onto pyrophyllite as a bidentate inner-sphere surface complex. They also suggest that Ni precipitates onto the pyrophyllite surface as a mixed Ni/Al hydroxy precipitate, especially above pH 7. This precipitation reaction occurs in a system that is undersaturated with respect to Ni. Cheah et al. (1998) found evidence of Cu(II) dimerization following inner-sphere complex formation between Cu and SiO_2.

The formation of precipitates in partitioning studies using metal salts is especially important. High concentrations of metal salts may lead to precipitation reactions in batch experiments that would not realistically be encountered in the field. High precipitation rates would lead to falsely high K_d values. Batch equilibrium experiments using metal salts should always use environmentally relevant concentrations (Hendrickson and Corey, 1981; Anderson and Christensen, 1988).

Laboratory Procedures that Affect K_d Values

Laboratory procedures that affect K_d values include: selection of digestion/extraction solutions to estimate $[M]_{soil}$ and $[M]_{solution}$, time of extraction or solution equilibration, method of metal equilibration with the soil, and ratio of extracting solution to soil. The effect of extraction type on K_d values has been addressed previously in this chapter and will not be expanded here.

The extraction time used in batch extractions typically used to calculate K_d values for soils is an important factor in partitioning experiments. Soils rarely, if ever, exist in a true state of chemical equilibrium (Sparks, 1995). Therefore, researchers depend on operationally defined equilibration times. A 24-h extraction time is typical for many batch extractions (Mitchell et al., 1978; Anderson and Christensen, 1988); 16-h extraction times are quite common in the literature also (Sposito et al., 1982; Miller et al., 1986); and extraction time may extend into several weeks (Jopony and Young, 1994). Time of extraction/equilibration will play a role in final K_d values for soils. Longer extraction times for unspiked soils will tend to increase metal in solution, thereby decreasing K_d. For experiments where soils are equilibrated with metal solutions, shorter equilibration times will tend to leave more metal in solution (especially for reactions with relatively slow kinetics), thereby decreasing K_d.

Experiments examining partitioning of metals in soils may fall into two broad categories: experiments where a metal salt solution is equilibrated for some time with soil (Lee et al., 1996) and experiments where an unamended soil is equilibrated with an extracting solution (Gooddy et al., 1995). Depending on the amount and nature of metal binding sites, the K_d values for a particular soil may differ when comparing results from both types of equilibration techniques. Any precipitation during equilibration will be interpreted as an addition to the $[M]_{soil}$ component of K_d, which will increase the K_d value. Though valid in the laboratory, a true field soil may never encounter the concentration of metal in a metal salt equilibration experiment. Unspiked soils that are extracted will tend to have metals that are much more difficult to extract and therefore have higher relative K_d values, but these values may be more applicable to field situations. Regardless of the type of equilibration, field conditions should be mimicked as closely as possible. When studying partitioning of metals, the researcher needs to identify the goal of the research. For example, if information on partitioning of metals from a spill is needed, a metal salt equilibration type experiment would be applicable. Conversely, if research on the effects of acid rain on partitioning of metals is desired, an equilibration with a dilute acid may be most suitable.

The ratio of soil to solution plays a significant role in the results of metal partitioning studies. K_d values decrease with increasing soil concentration. This has been described as the solids effect (O'Conner and Connolly, 1980; Voice et al., 1983; Di Toro, 1985; Celorie et al., 1989). Grover and Hance (1970) suggested that this effect is predominantly caused by higher surface area exposure at low soil:solution ratios. The low ratios allow relatively greater sorption of metals at low soil solution ratios, and therefore higher K_d values. Another explanation is that there are simply more particles that pass through a given filter at higher solids concentrations. More particles transporting bound metal through a filter are analyzed as "soluble" or desorbed metals in a supernatant yielding lower K_d values (Voice et al., 1983; Voice and Weber, 1985; Van Benschoten et al., 1998). Data from this laboratory (presented later in this chapter) offer compelling evidence that supports the concept of increased unfilterable particles in higher soil:solution extractions causes K_d values to be lowered. Similarly, the effects of shaking rates for batch extractions may contribute to the solids effect. Higher rates of agitation will tend to increase the amount of small particles due to particle-particle interaction and abrasion (Sparks, 1995). Celorie et al. (1989) suggested the use of a centrifugation technique in place of batch extractions to eliminate solids effects.

When using batch extraction/equilibration techniques, natural conditions should be reproduced as closely as possible. The soil:solution ratio should be maximized while remaining operational. It is also essential for researchers to define the focus of the partitioning studies. For example, if the goal is to analyze potentially partitionable metal in a soil system that could enter an aquifer, then all forms of metal passing through a particular filter may be considered soluble regardless of whether or not they are bound to a colloid or are truly soluble.

EFFECTS OF SOIL PARAMETERS AND OPERATIONAL PROCEDURE ON METAL PARTITIONING: A CASE STUDY

To illustrate the importance of soil parameters and soil:solution ratio on metal partitioning, we studied desorption of three metals, Cu, Ni, and Zn, from 15 soils. The major soil parameters responsible for desorption of these metals from soils were elucidated. Models were developed to predict the partitioning of metals to soil and the aqueous speciation.

Fifteen New Jersey soils with texture ranging from sand to loam and organic C content from 1.2 to 49.9 g/kg were employed to conduct the experiments. The soil samples were air-dried and sieved through a 2-mm screen before use. Detailed characteristics of the soils have been reported by Yin et al. (2000). The total concentrations of metals in soils were determined by acid digestion following the U.S. EPA SW-846 method (USEPA, 1995). Adsorption of cadmium from a $1 \times 10^{-5}\,M$ solution was conducted at a soil:solution ratio of 1 g per 100 mL. Desorption of metals from soils was initialized by mixing each soil with deionized water at natural soil pH with no chemical amendments to the soils. The soil:DI H_2O extract ratio was 1 g:0.8 mL. This was the lowest operationally feasible ratio and was employed to closely mimic natural field conditions. The soil mixtures were equilibrated by shaking on a reciprocal shaker at 100 strokes per minute for 24 h at $25 \pm 1°C$. After equilibration, soil solids were separated from solution by centrifugation followed by filtration through a 0.45 μm pore size membrane filter. The final pH for each filtrate was determined by an Orion pH electrode. The concentrations of soluble metals and dissolved organic C in the filtrates were determined by a Spectro ICP and a Dohrmann DC-90 TOC analyzer, respectively. The free Cu^{2+} activities in the filtrates were determined by a Cu ion selective electrode.

The importance of soil organic matter in metal partitioning was emphasized by Lee et al. (1996) who demonstrated that the relationship between log K_d and pH for the adsorption of cadmium was improved by almost one order of magnitude by normalizing the partitioning coefficient to the amount of organic matter present in the soil as shown in Figure 7.1. When K_{om} rather than K_d is considered, the R^2 value increases from 0.799 to 0.927. Deviation of values from the line shown in Figure 7.1b are a result of the effects of desorption of organic matter at high pH and of dissolution

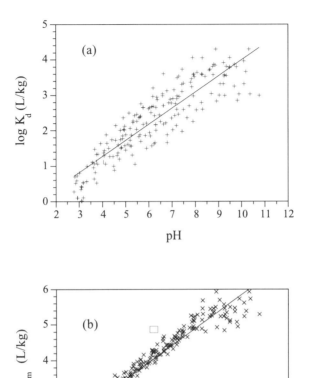

Figure 7.1 Adsorption of cadmium onto 15 New Jersey soils adjusted to various pH values expressed as (a) K_d and (b) K_{om}. (Adapted with permission from Lee, S.Z., H.E. Allen, C.P. Huang, D.L. Sparks, P.F. Sanders, and W.J.G.M. Peijnenburg. 1996. Predicting soil-water partitioning coefficients for cadmium. *Env. Sci. Tech.* 30:3418-3424. ©1996. American Chemical Society.)

of aluminum and other metals at low pH. This approach differs from that usually used in determining the soil components important to partitioning, multiple linear regression analysis. In multiple linear regression analyses, pH and the concentrations of organic matter and metal oxide solid phases are treated as if they are independent parameters. However, the strength of sorption of cadmium and other metals to the solid phases is not independent of the solution pH. The strength and capacity of metal oxides and organic matter to bind metals increases with increasing pH (Lion et al., 1982).

We found that increases in the soil:water ratio decreased K_d for copper and organic matter, as is shown in Figure 7.2. K_d values for metals were determined from

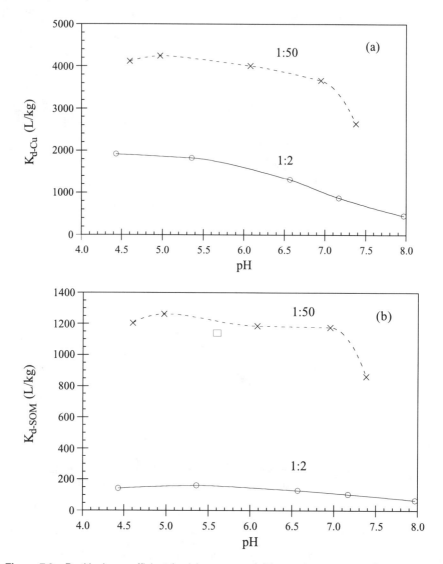

Figure 7.2 Partitioning coefficient for (a) copper and (b) organic matter as a function of pH for Washington loam as a function of pH at 1:2 and 1:50 soil:solution ratios (g soil:mL solution). (Adapted from Yin, Y., C.A. Impellitteri, S.J. You, and H.E. Allen. 2000. The importance of organic matter distribution and extract soil:solution ratio on the desorption of heavy metals from soils. *Sci. Tot. Env.* (Submitted).)

the measurements of total metal in the soil (C_T, mg/kg) and metal in the soil solution (C_w, mg/L). The partitioning coefficient for the soil organic matter was calculated in a similar manner. To mimic field conditions, we focused on studies at a minimum soil/water ratio of 25 g/20 mL, which was close to field capacity and was also lab operable. The results indicate that soil pH and organic matter are the major parameters determining the partitioning of these three metals. Among the three metals studied, Cu has strong binding affinity for both soluble and particulate organic matter.

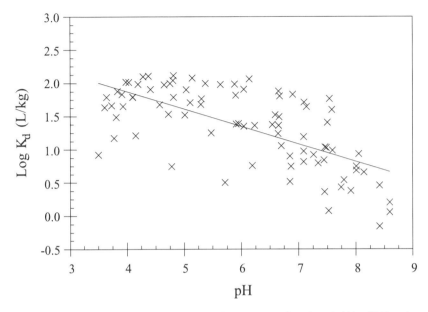

Figure 7.3 Partitioning coefficient for soil organic matter as a function of pH for 15 New Jersey soils adjusted to various pH values. (Adapted from You, S.J., Y. Yin, and H.E. Allen. 1999. Partitioning of organic matter in soils: effects of pH and water/soil ratio. *Sci. Tot. Env.* 227:155-160. With permission from Elsevier Science.)

Consequently, soil-water partitioning of Cu largely depended on the partitioning of organic matter between soil solids and solution phases.

The partitioning of the organic matter between the soil and the soil solution (Equation 7.10) is very dependent on the pH of the solution. Figure 7.3 shows the variation of the partitioning coefficients for organic matter in the 15 New Jersey soils. Subsamples of each soil were adjusted in pH to cover the range of approximately 3.5 to 8.5. The correlation coefficient, R^2, for the regression is 0.45. We have also shown (You et al., 1999) that not only is more organic matter released to the solution as the pH is raised, but that the ratio of humic to fulvic acid in the solution increases with increasing pH. This has significant consequences to the speciation of metals because humic acid binds metals such as copper much more strongly than fulvic acid. Computations using WHAM (Tipping, 1994) indicate that at a constant amount of organic matter the free copper ion could decrease by more than an order of magnitude at pH 6 as the humic acid proportion increased from 10 to 20% of the organic matter.

Among the three metals studied, Cu has the strongest binding affinity for both soluble and particulate organic matter. Consequently, soil-water partitioning of Cu largely depends on the partitioning of organic matter between soil solids and solution phases (Yin et al., 2000). The partitioning coefficient for soluble Cu correlated well with the partitioning coefficient for organic matter with a regression coefficient R^2 of 0.91 (Figure 7.4). The partitioning of free Cu^{2+} ion, however, was highly pH dependent and also closely related to the total soil organic C content. Normalization of the partition coefficient for free Cu^{2+} (calculated as the ratio of Cu adsorbed per

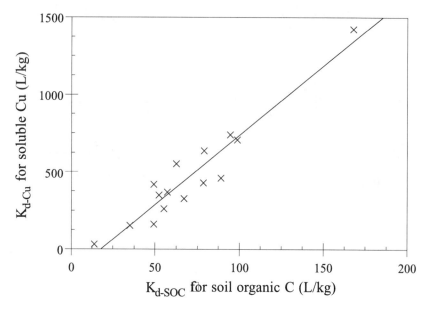

Figure 7.4   Correlation between partitioning coefficient (l/kg) for soluble Cu (K_{d-Cu}) and that
for soil organic C (SOC) (K_{d-SOC}) at natural soil pH. (Adapted from Yin, Y., C.A.
Impellitteri, S.J. You, and H.E. Allen. 2000. The importance of organic matter
distribution and extract soil:solution ratio on the desorption of heavy metals from
soils. *Sci. Tot. Env.* (Submitted).)

mass of soil to the concentration of free Cu^{2+} ion) with the soil organic C content
significantly improved the correlation between the partition coefficient for free Cu^{2+}
ion and pH, with the regression coefficient R^2 increasing from 0.80 to 0.96.

Unlike Cu, the effect of dissolved organic matter on the partitioning of Ni and
Zn was not important compared with that of particulate organic matter at the natural
soil pH. Complexation of zinc and nickel with dissolved organic matter is much
weaker than is that of copper. The partitioning coefficient increased with soil pH
for both metals. Yet the partitioning of both metals was affected by soil organic C
content. For these results, normalization of the partitioning coefficient with soil
organic C content also significantly improved correlations between the normalized
partitioning coefficient and pH for both metals with the regression coefficient increasing from 0.53 to 0.87 for Ni and from 0.62 to 0.83 for Zn. The improvements in
the relationships suggest that soil organic matter is the dominant soil constituent
responsible for the partitioning of these metals.

SUMMARY AND CONCLUSIONS

Soil pH and organic matter are the two most important parameters determining
metal partitioning and aqueous speciation in soils. By incorporating both soluble
and particulate organic matter and considering the effect of pH independently,
correlations can be developed to predict metal partitioning and speciation in soils.

However, we have also shown the importance of the soil:solution ratio in affecting the measured partitioning coefficient. Caution must be exercised in the use of laboratory measurements of K_d for the prediction of metal partitioning in field situations where the ratio can be very different.

REFERENCES

Ainsworth, C.C., J.L. Pilon, P.L. Gassman, and W.G.V.D. Sluys. 1994. Cobalt, cadmium, and lead sorption to hydrous iron oxide: residence time effect. *Soil Sci. Soc. Am. J.* 58:1615-1623.

Allen, H.E. and Y. Yin. 1996. The importance of organic matter on the sorption of cadmium and mercury to soil. The 6th International Conference on Preservation of Our World in the Wake of Change, Jerusalem, Israel.

Anderson, P.R. and T.H. Christensen. 1988. Distribution coefficients of Cd, Co, Ni, and Zn in soils. *J. Soil Sci.* 39:15-22.

Backes, C.A., R.G. McLaren, A.W. Rate, and R.S. Swift. 1995. Kinetics of cadmium and cobalt desorption from iron and manganese oxides. *Soil Sci. Soc. Am. J.* 59:778-785.

Barrow, N.J. 1986. Testing a mechanistic model. IV. Describing the effects of pH on zinc retention by soils. *J. Soil Sci.* 37:295-303.

Bunzl, K., W. Schmidt, and B. Sansoni. 1976. Kinetics of ion exchange in soil organic matter. IV. Adsorption and desorption of Pb^{2+}, Cu^{2+}, Cd^{2+}, Zn^{2+}, and Ca^{2+} by peat. *J. Soil Sci.* 27:32-41.

Celorie, J.A., S.L. Woods, T.S. Vinson, and J.D. Istok. 1989. A comparison of sorption equilibrium distribution coefficients using batch and centrifugation methods. *J. Env. Qual.* 18:307-313.

Cheah, S.F., J.G.E. Brown, and G.A. Parks. 1998. XAFS spectroscopy study of Cu(II) sorption on amorphous SiO_2 and g-Al_2O_3: effect of substrate and time on sorption complexes. *J. Colloid Interface Sci.* 208:110-198.

Di Toro, D.M. 1985. A particle interaction model of reversible organic chemical sorption. *Chemosphere.* 14:1503-1538.

Di Toro, D.M., J.D. Mahony, P.R. Kirchgraber, A.L. O'Byrne, L.R. Pasquale, and D.C. Piccirilli. 1986. Effects of nonreversibility, particle concentration, and ionic strength on heavy metal sorption. *Env. Sci. Tech.* 20:55-61.

Egozy, Y. 1980. Adsorption of cadmium and cobalt on montmorillonite as a function of solution composition. *Clays Clay Miner.* 28:311-318.

Elliott, H.A., M.R. Liberati, and C.P. Huang. 1986. Competitive adsorption of heavy metals by soils. *J. Env. Qual.* 15:214-219.

Evans, L.J. 1989. Chemistry of metal retention by soils. *Env. Sci. Tech.* 23:1046-1056.

Farrah, H. and W.F. Pickering. 1976. The sorption of copper species by clays. I Kaolinite. *Aust. J. Chem.* 29:1167-1176.

Ford, R.G., P.M. Bertsch, and K.J. Farley. 1997. Changes in transition and heavy metal partitioning during hydrous iron oxide aging. *Env. Sci. Tech.* 31:2028-2033.

Gerritse, R.G. and W.V. Driel. 1984. The relationship between adsorption of trace metals, organic matter, and pH in temperate soils. *J. Env. Qual.* 13:197-204.

Gooddy, D.C., P. Shand, D.G. Kinniburgh, and W.H.v. Riemsdijk. 1995. Field-based partition coefficients for trace elements in soil solutions. *Eur. J. Soil Sci.* 46:265-285.

Grover, R. and R.J. Hance. 1970. Effect of ratio of soil to water on adsorption of linuron and atrazine. *Soil Sci.* 100:136-138.

Harter, R.D. 1983. Effect of soil pH on adsorption of lead, copper, zinc, and nickel. *Soil Sci. Soc. Am. J.* 47:47-51.

Hendrickson, L.L. and R.B. Corey. 1981. Effect of equilibrium metal concentrations on apparent selectivity coefficients of soil complexes. *Soil Sci.* 131:163-171.

Hogg, D.S., R.G. McLaren, and R.S. Swift. 1993. Desorption of copper from some New Zealand soils. *Soil Sci. Soc. Am. J.* 57:361-366.

Janssen, R.P.T., W.J.G.M. Peijnenburg, L. Posthuma, and M.A.G.T.v.d. Hoop. 1997. Equilibrium partitioning of heavy metals in Dutch field soils. I. Relationship between metal partition coefficients and soil characteristics. *Env. Tox. Chem.* 16:2470-2478.

Janssen, R.P.T., L. Posthuma, R. Baerselman, H.A.D. Hollander, R.P.M.V. Veen, and W.J.G.M. Peijnenburg. 1997. Equilibrium partitioning of heavy metals in dutch field soils. II. Prediction of metal accumulation in earthworms. *Env. Tox. Chem.* 16:2479-2488.

Jopony, M. and S.D. Young. 1994. The solid-solution equilibria of lead and cadmium in polluted soils. *Eur. J. Soil Sci.* 45:59-70.

Lee, S.Z., H.E. Allen, C.P. Huang, D.L. Sparks, P.F. Sanders, and W.J.G.M. Peijnenburg. 1996. Predicting soil-water partitioning coefficients for cadmium. *Env. Sci. Tech.* 30:3418-3424.

Lion, L.W., R.S. Altmann, and J.O. Leckie. 1982. Trace metal adsorption characteristics of estuarine particulate matter: evaluation of contributions of Fe/Mn oxide and organic surface coatings. *Env. Sci. Tech.* 16:660-666.

McBride, M., S. Sauvé, and W. Hendershot. 1997. Solubility control of Cu, Zn, Cd and Pb in contaminated soils. *Eur. J. Soil Sci.* 48:337-346.

McBride, M.B. 1994. *Environmental Chemistry of Soils.* New York, NY, Oxford University Press, Inc.

Miller, W.P., D.C. Martens, and L.W. Zelazny. 1986. Effect of sequence in extraction of trace metals from soils. *Soil Sci. Soc. Am. J.* 50:598-601.

Mitchell, G.A., F.T. Bingham, and A.L. Page. 1978. Yield and metal composition of lettuce and wheat grown on soils amended with sewage sludge enriched with cadmium, copper, nickel, and zinc. *J. Env. Qual.* 7:165-171.

O'Conner, D.J. and J.P. Connolly. 1980. The effect of concentration of adsorbing solids on the partition coefficient. *Wat. Res.* 14:1517-1523.

Sauvé, S., W. Hendershot, and H.E. Allen. 2000. Solid-solution partitioning of metals in contaminated soils: dependence on pH, total metal burden, and organic matter. *Env. Sci. Tech.* 34:1125-1131.

Sauvé, S., M. McBride, and W. Hendershot. 1998. Soil solution speciation of lead (II): effects of organic matter and pH. *Soil Sci. Soc. Am. J.* 62:618-621.

Sauvé, S., M.B. McBride, W.A. Norvell, and W.H. Hendershot. 1997. Copper solubility and speciation of *in situ* contaminated soils: effects of copper level, pH, and organic matter. *Wat. Air Soil Poll.* 100:133-149.

Scheidegger, A.M., M. Fendorf, and D.L. Sparks. 1996. Mechanisms of nickel sorption on pyrophyllite: macroscopic and microscopic approaches. *Soil Sci. Soc. Am. J.* 60:1763-1772.

Sparks, D.L. 1995. *Environmental Soil Chemistry.* San Diego, CA, Academic Press, Inc.

Sposito, G., L.J. Lund, and A.C. Chang. 1982. Trace metal chemistry in arid-zone field soils amended with sewage sludge: I. Fractionation of Ni, Cu, Zn, Cd, and Pb in solid phases. *Soil Sci. Soc. Am. J.* 46:260-264.

Temminghoff, E.J.M., S.E.A.T.M.v.d. Zee, and M.G. Keizer. 1994. The influence of pH on the desorption and speciation of copper in a sandy soil. *Soil Sci.* 158:398-408.

Tipping, E. 1994. WHAM-A chemical equilibrium model and computer code for waters, sediments, and soils incorporating a discrete site/electrostatic model of ion-binding by humic substances. *Computers Geosci.* 20:973-1023.

USEPA. 1995. *Test Methods for Evaluating Solid Waste. Vol. IA: Laboratory Manual Physical/Chemical Methods,* SW 846, 3rd ed. Washington, D.C., U.S. Government Printing Office.

USEPA. 1997. *Method 3051: Microwave Assisted Acid Dissolution of Sediments, Sludges, Soils, and Oils.* 2nd ed. Washington, D.C., U.S. Government Printing Office.

Van Benschoten, J.E., W.H. Young, M.R. Matsumoto, and B.E. Reed. 1998. A nonelectrostatic surface complexation model for lead sorption on soils and mineral surfaces. *J. Env. Qual.* 27:24-30.

Voice, T.C., C.P. Rice, and J. W.J. Weber. 1983. Effect of solids concentration on the sorptive partitioning of hydrophobic pollutants in aquatic systems. *Env. Sci. Tech.* 14:513-518.

Voice, T.C. and W.J. Weber. 1985. Sorbent concentration effects in liquid/solid partitioning. *Env. Sci. Tech.* 19:789-796.

Yin, Y., C.A. Impellitteri, S.J. You, and H.E. Allen. 2000. The importance of organic matter distribution and extract soil:solution ratio on the desorption of heavy metals from soils. *Sci. Tot. Env.* (Submitted).

You, S.J., Y. Yin, and H.E. Allen. 1999. Partitioning of organic matter in soils: effects of pH and water/soil ratio. *Sci. Tot. Env.* 227:155-160.

CHAPTER 8

Understanding Sulfate Adsorption Mechanisms on Iron (III) Oxides and Hydroxides: Results from ATR-FTIR Spectroscopy

Derek Peak, Evert J. Elzinga, and Donald L. Sparks

INTRODUCTION

Sulfate ($SO_{4\,aq}^{2-}$) is a weakly basic Group VI oxyanion with a metal center that has a charge of +6. In aqueous solution it exists as either the fully-deprotonated form, or as the singly protonated bisulfate ($HSO_{4\,aq}^{-}$) ion (Stumm and Morgan, 1991). The pK_a for the protonation reaction is ~1.9, making the fully deprotonated form the dominant ion under normal soil conditions. Sulfate ions have a hydrated radius of about 4 Å. At the present time, the chemistry of sulfate in the soil environment is still poorly understood. In fact, the mechanisms of sulfate sorption have often been the subject of debate, both historically and in the current scientific literature. Sulfate is of interest to soil chemists for environmental and agronomic reasons. It is an essential micronutrient for plant growth. Neither deficiency nor toxicity symptoms are commonly seen in cultivated soils, but sulfate can occur in extremely high levels near sites of mine waste deposition as a result of hydrogen sulfide oxidation (Persson and Lövgren, 1996). Sulfate is a product in the geochemical cycling of pyrite and therefore plays an important role in marine sediment chemistry.

Macroscopic studies of sulfate sorption have suggested that sulfate adsorbs via an outer-sphere (electrostatic) adsorption mechanism on both soils and reference minerals (Charlet et al., 1993). This conclusion is supported primarily by two observations: (1) ionic strength has a great effect on the amount of sulfate that is adsorbed, with increasing adsorption as ionic strength decreases, and (2) no adsorption of sulfate is usually seen above the point of zero charge of the mineral. This fact potentially makes iron and aluminum oxides important sites for sulfate adsorption in soils, since these components have high points of zero charge and are commonly found in soils.

He and colleagues (1996) studied the stoichiometry of hydroxyl-sulfate exchange on gamma aluminum oxide and kaolinite using a back-titration technique. Combining a thermodynamic approach with their data, they suggested that the observed hydroxyl release upon sulfate adsorption need not be associated with a ligand-exchange mechanism. Instead, they suggested a mechanism consisting initially of surface site protonation and generation of a hydroxyl in solution via a reaction such as $Al–OH^0 + H_2O \rightarrow Al–OH_2^+ + OH^-$. They proposed that this protonation was then followed by the formation of an outer-sphere surface complex: $Al–OH_2^+ SO_4^{2-}$. This mechanism accounts for the observed proton consumption to maintain pH (neutralizing the hydroxyl generated) without requiring an inner-sphere surface complex.

It has also been shown that the rate of gibbsite dissolution in the presence of sulfate is more rapid than in the presence of chloride (Ridley et al., 1997). While an enhancement of dissolution in the presence of ligands is often attributed to inner-sphere surface complex formation, the explanation for this observed effect was the formation of aluminum-sulfate complexes in solution that enhance mineral solubility. These aluminum-sulfate complexes result in more aluminum being released from the surface because they keep the free aluminum concentration in solution much lower than when only chloride is present.

Sposito (1984) suggested that sulfate adsorption might be of an intermediate nature, sometimes sorbing as an outer-sphere complex and sometimes as an inner-sphere complex via a ligand exchange mechanism. This concept was supported by the observations of Yates and Healy (1975), who investigated sulfate adsorption on both α-FeOOH and α-Cr$_2$O$_3$. Although the rates of hydroxyl exchange for the two sorbents are markedly different, the rate and extent of sulfate adsorption was very similar, implying an outer-sphere complexation mechanism. However, sulfate adsorption also shifts the point of zero charge to higher values on both α-FeOOH and α-Cr$_2$O$_3$, which is consistent with inner-sphere complexation.

Many research groups have modeled sulfate adsorption on both soils and pure mineral components, with differing results. He and co-workers (1997) utilized the triple layer model with outer-sphere surface complex formation to describe sulfate adsorption on γ-alumina and kaolinite. Charlet and co-workers (1993) modeled the adsorption of sulfate on an aluminum-coated TiO$_2$, δ-Al$_2$O$_3$, and an acidic forest soil. They found that outer-sphere complexation described the data well. Sulfate adsorption on goethite has been modeled using several different approaches. Zhang and Sparks (1990) modeled sulfate adsorption on goethite using the triple-layer model and an outer-sphere surface complex. Davis and Leckie (1980) employed a modified triple-layer model to represent sulfate adsorption on goethite. They determined that a mixture of outer-sphere and inner-sphere sulfate surface complexes best described their data. The charge-distribution multisite complexation approach (CD-Music) has been used by several researchers to model sulfate adsorption on goethite. Geelhoed and colleagues (1997) described sulfate complexation using an inner-sphere bidentate binuclear surface complex, while Rietra and co-workers (1999) used an inner-sphere monodentate surface complex. Persson and Lövgren

(1996) utilized an extended double-layer model to describe their experimental data and concluded that an outer-sphere surface complex was most likely. Ali and Dzombak (1996) modeled adsorption of sulfate on goethite in both the absence and presence of simple organic acids using a generalized two-layer model. Three different inner-sphere surface complexes of sulfate were required to describe the experimental data.

There are a few possible reasons that modelers have not completely agreed about the nature of sulfate-goethite complexation mechanisms. First of all, different synthesis methods and pretreatment techniques can result in goethite with quite different surface properties. If the morphology and crystallinity of the sorbent varies, then the surface chemistry can sometimes also be different. For example, Sujimoto and Wang (1998) found that hematite morphology had a significant effect on the mechanism of sulfate sorption. Another possible source of error is the fact that the design of some surface complexation models excludes some potential surface complexes. This could result in an incorrect assignment of sulfate surface-complexation mechanisms. For example, the original triple-layer model only considered outer-sphere complex formation, and the CD-MUSIC model as currently implemented only describes inner-sphere complex formation. A third possibility is that none of the surface complexation models used are robust enough to converge only on one unique solution when several surface complexes are occurring simultaneously. Definitive mechanistic information from spectroscopy can constrain models to physically relevant complexes and can therefore improve model refinement.

Although somewhat contradictory to the macroscopic laboratory studies, there is microscopic and spectroscopic evidence of sulfate inner-sphere surface complexation. Transmission infrared spectroscopic studies of sulfate adsorption on goethite and hematite (Parfitt and Smart, 1978; Turner and Kramer, 1991) revealed the formation of sulfate bidentate binuclear surface complexes on both solids. XPS studies (Martin and Smart, 1987) also validated this sorption mechanism. More recently, Persson and Lövgren (1996) concluded that outer-sphere adsorption of sulfate on goethite was occurring based on results from diffuse reflectance infrared (DRIFT) spectroscopy. However, these spectroscopic experiments all involved potential sample alteration via either drying, the application of heat and pressure, or dilution in a salt, which could have modified the structure of the original sorption complex.

Due to the potential artifacts in *ex situ* spectroscopy, it is greatly preferable to conduct *in situ* experimentation to elucidate interactions that occur in aqueous suspensions. *In situ* experiments using scanning tunneling microscopy (STM) (Eggleston et al., 1998) and ATR-FTIR spectroscopy (Eggleston et al., 1998; Hug, 1997) to determine the adsorption mechanism of sulfate on hematite have more recently shown that inner-sphere monodentate surface complexes form at the hematite surface under aqueous conditions. Degenhardt and McQuillan (1999) found that sulfate forms primarily outer-sphere surface complexes on chromium (III) oxide hydroxide, with some splitting of infrared bands being observed and attributed to electrostatic forces. Peak et al. (1999) utilized ATR-FTIR to better understand the adsorption of sulfate on goethite. We determined that sulfate forms only outer-sphere surface complexes above pH 6.0, and that it forms a mixture of outer- and inner-sphere surface complexes at pH less than 6.0. This is quite significant, as it demonstrates a continuum between different adsorption mechanisms that can potentially explain the discrepancies in earlier macroscopic and surface complexation modeling research.

Hodges and Johnson (1987) used a miscible displacement technique to follow the kinetics of sulfate adsorption/desorption on goethite. They found that the reaction kinetics could best be described using diffusion-limited kinetics models. This is not surprising considering the nature of the miscible displacement technique. A miscible displacement setup is a continuous flow system where a small amount of sorbent is injected into a thin disk (usually a filter holder with filter paper). A dilute solution of the sorbate in a constant ionic strength background is flowed through this disk until equilibrium is reached. At this point the solution is changed to pure background electrolyte, and the desorption reaction is monitored. The main problems with this experiment are that there is little mixing inside the thin disk and it is difficult to ensure that the sorbent is distributed evenly inside the thin disk without preferential flow pathways. Both these drawbacks tend to make diffusional forces extremely important in modeling the experimental results. Hodges and Johnson (1987) were unable to clearly determine whether sulfate adsorption was due to ligand exchange or electrostatic attraction in these studies. Pressure-jump chemical relaxation studies investigating sulfate adsorption on goethite (Zhang and Sparks, 1990) suggested an outer-sphere complexation mechanism. This seems somewhat at odds with the recent observation that sulfate forms both outer-sphere and inner sphere complexes on goethite at pH below 6. However, pressure-jump studies cannot conclusively determine the mechanism by which a reaction proceeds. Instead they determine the number of reaction steps and the rate constants for those individual steps. This information must then be coupled with an equilibrium model and/or spectroscopic studies to accurately assess the identity of these individual reaction products. Furthermore, if reactions are occurring simultaneously rather than sequentially, then a pressure-jump experiment may not distinguish them effectively if the time scales of the reactions are similar.

The relationship between the symmetry of sulfate complexes and their infrared spectra is well established (Nakamoto, 1986), and it is possible to assign molecular symmetry based on the number and position of peaks that appear in the mid-infrared region. The relationship between the symmetry of surface complexes and the resulting infrared spectrum is summarized in Figure 8.1. With the Attenuated Total Reflectance (ATR) technique under aqueous conditions, there are two infrared sulfate vibrations that are accessible to spectroscopic investigation. They are the non-degenerate symmetric stretching v_1, and the triply degenerate asymmetric stretching v_3 bands (Persson and Lövgren, 1996). As a free anion in solution, sulfate has tetrahedral symmetry and belongs to the point group T_d. For this symmetry, only one broad peak at approximately 1100 cm^{-1}, due to the triply degenerate v_3 band, is usually observed. In some cases the v_1 band is also weakly active and appears at around 975 cm^{-1}. Since outer-sphere complexes retain their waters of hydration and form no chemical bonds, it is expected that the symmetry of outer-sphere sulfate

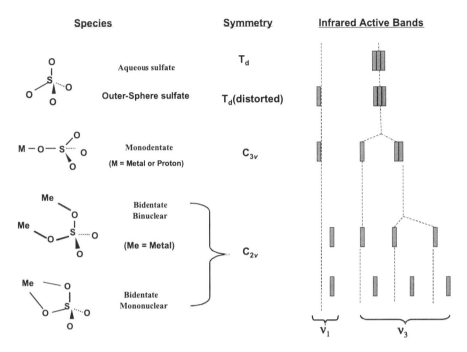

| Species | Symmetry | Infrared Active Bands |

Figure 8.1   The relationship between the molecular symmetry of sulfate complexes and the observed infrared spectrum they produce. (Adapted from Hug, S. J. (1997). *In situ* Fourier transform infrared measurements of sulfate adsorption on hematite in aqueous solutions. *J. Colloid Interface Sci.* 188: 415-422. With permission.)

complexes is similar to aqueous sulfate. However, distortion due to electrostatic effects could shift the v_3 to higher wave number and cause the v_1 band to become IR active. If sulfate is present as an inner-sphere complex, the symmetry is lowered. As a result, the v_1 band becomes infrared active, and the v_3 band splits into more than one peak. In the case of a monodentate inner-sphere surface complex, such as observed for sulfate adsorption on hematite, C_{3v} symmetry results. The v_3 band splits into two peaks, one at higher wave number and one at lower wave number, while the v_1 band becomes fully active at about 975 cm^{-1} (Hug, 1997; Nakamoto, 1986). If sulfate forms a bidentate binuclear (bridging) surface complex, the symmetry is further lowered to C_{2v}, and the v_3 band splits into three bands between 1050 and 1250 cm^{-1}, while the v_1 band is shifted to around 1000 cm^{-1} (Nakamoto, 1986). Figure 8.2 contains a table (adapted from Hug, 1997) that details the positions of v_1 and v_3 sulfate bands reported in the literature for different molecular configurations.

In this chapter, we present results from investigation of the mechanism of sulfate adsorption on goethite, hematite, and ferrihydrite as a function of pH. Additionally, the effects of surface loading and ionic strength on sulfate adsorption on goethite were studied. Understanding how adsorption mechanisms are affected by reaction conditions is of considerable interest to developers of surface complexation models as well as soil scientists, but there are few studies that use *in situ* spectroscopy to study soil chemical reactions over a wide range of conditions.

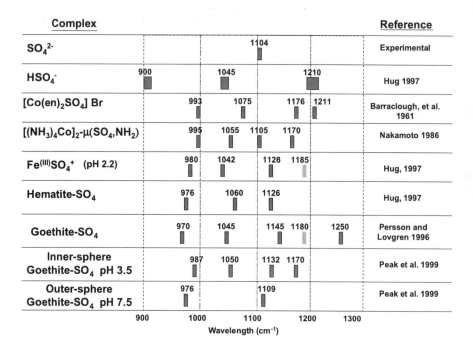

Figure 8.2 Table of infrared peak positions for relevant sulfate reference compounds in the literature. (Adapted from Hug, S. J. (1997). *In situ* Fourier transform infrared measurements of sulfate adsorption on hematite in aqueous solutions. *J. Colloid Interface Sci.* 188: 415-422. With permission.)

APPROACHES

Mineral Synthesis

The goethite used in this study was synthesized using the method of Schwertmann et al. (1985). Initially, ferrihydrite was precipitated by adding 50 mL of 1 M ferric nitrate solution to 450 mL of 1 M KOH. This suspension of amorphous hydrous ferric oxide was then aged for 14 days at 25° C. The suspension was next washed via centrifugation, replacing the supernatant with doubly deionized water to remove residual KOH. The rinsed solid was then resuspended in 0.4 M HCl and shaken for 2 hours using a mechanical shaker. This treatment was used to remove any remaining ferrihydrite from the surface of the goethite. The acidified goethite suspension was again washed via centrifugation to remove both HCl and dissolved iron. Finally, the goethite was dialyzed, frozen with liquid nitrogen, and freeze-dried. The solid was confirmed as goethite via infrared spectroscopy using both ATR and transmission-mode KBr pellets. The external surface area determined from N_2 BET was 63.5 m^2g^{-1}, and the point of zero salt effect was 8.4, as determined via potentiometric titration in 0.1, 0.01, and 0.005 M sodium perchlorate. The hematite used in these experiments was synthesized from ferric perchlorate using the method of Schwertmann and Cornell (1991). It was also acid-washed and dialyzed prior to freeze-drying, and had an N_2-BET surface area of 14 m^2g^{-1}. Ferrihydrite (6 line) was synthesized by titrating 1 M ferric chloride to pH 7.5 with 1 M KOH. This precipitate was washed 3 times with 0.1 M NaCl to remove any residual iron, washed once with deionized water, and then dialyzed for 3 days in deionized water.

Attenuated Total Reflectance (ATR-FTIR) Spectroscopy

Attenuated Total Reflectance Fourier Transform Infrared (ATR-FTIR) spectros-copy is ideally suited to probe sulfate adsorption mechanisms on iron oxides since there are no overlapping bands in the mid-infrared region. As described earlier, the relationship between symmetry and infrared spectra is well established for sulfate, so it is possible to identify surface complexes based on the number and position of peaks in the mid-IR range.

All FTIR experiments were conducted using a Perkin-Elmer 1720x spectrometer equipped with a purge gas generator and either a DTGS or MCT detector. A horizontal ATR accessory and flow cell (Spectra-Tech) were used for sampling. A flow-through system has been developed by our group (Peak et al., 1999) that greatly improves the applicability of ATR-FTIR spectroscopy for the study of interfacial chemistry. The sorbent phase was first deposited onto the 45° zinc selenide crystal using a modification of the method of Hug (1997). Once dry, the deposited metal oxide was rinsed with 0.01 N NaCl to remove any nonadhering sorbent and allowed to air-dry again. The crystal was then placed into a flow cell. Once the crystal containing the deposited sorbent is placed into the flow cell, Tygon tubing connects the flow cell to a peristaltic pump that delivers solution at a constant flow rate from a reaction vessel containing water, sulfate, and an inert salt (NaCl) to maintain constant ionic strength. The reaction vessel was nitrogen purged and pH controlled. It is possible to conduct adsorption edges, adsorption isotherms, and kinetics exper-iments directly within the flow cell using this experimental setup. Figure 8.3 shows a schematic of the flow cell experimental setup.

The greatest advantage of this flow-through type experimental setup is that aqueous sulfate concentrations remain low throughout the experiment. In fact, even in isotherms where aqueous sulfate concentrations reached 500 μM, the absorbance accounts for less than 0.1% of the total infrared absorbance in the flow cell. The deposited sorbent greatly concentrates the adsorbed sulfate in the path of the IR beam because a large volume of low concentration reactant flows through it during the experiment. It is therefore possible to avoid many spectral subtractions of supernatants that are common when using pastes to analyze sorption samples with ATR-FTIR spectroscopy. This can be especially useful when outer-sphere complexation is occurring, because an outer-sphere complex may be difficult to separate from the nonadsorbing aqueous reactant. Another advantage is that a better subtraction of the background (sorbent and background electrolyte) is possible, and this leads to better data quality over a wider spectral range.

pH Envelopes

To generate pH envelopes, the outflow tube from the flow cell was connected to the reaction vessel containing background electrolyte solution at a pH where adsorption

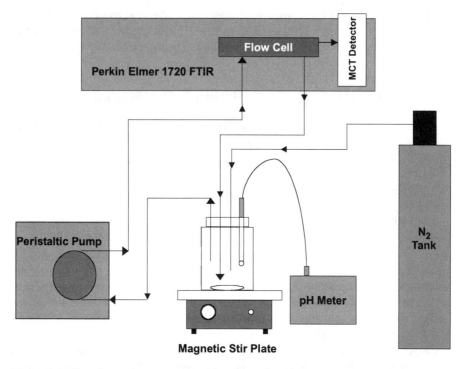

Figure 8.3 Experimental apparatus used for pH envelopes and adsorption isotherms.

of the sulfate is very low (pH 9.0). Once a stable background was collected, sulfate was then added, and spectra were collected until there was no increase in the intensity of the spectra with time. This is operationally defined as an equilibrium state. At this point, the pH was lowered so more adsorption occurs. Spectra were collected in this manner from pH 9 to pH 3.5.

Adsorption Isotherms

For the adsorption isotherms, the effluent from the flow cell is collected as waste instead of being circulated back to the reaction vessel. This was done to ensure that the equilibrium sulfate concentration (C_{eq}) can be determined. When the sulfate initially enters the flow cell, it is rapidly adsorbed to the deposited goethite in the flow cell, and the sulfate concentration in the effluent will remain low. As the system approaches equilibrium, however, the effluent concentration rises until it equals the influent concentration. At this time, the influent concentration is equal to C_{eq}, and the system is at equilibrium.

For these experiments, equilibrium is again defined as the point where no further increase in the infrared spectra of the adsorbed reactant is observed. The amount of reactant needed to adjust the remaining volume to the next concentration can then be calculated using the flow rate of the pump and the time elapsed since the pump was started. The reactant concentration in the reaction vessel is then raised and allowed to reach a new equilibrium with the sorbent in the flow cell. This is repeated to generate spectra of adsorbed reactant as a function of equilibrium reactant concentration. The integrated absorbance of these samples is then plotted vs. C_{eq}, as in a traditional isotherm.

Data Analysis

Peak Solve for Windows (Galactus Industries) was used to fit a linear baseline to all spectra and also to fit peaks to these corrected spectra. It was determined that Gaussian peaks best described all of the raw sulfate spectra as well as the aqueous reference samples. When inner-sphere sulfate surface complexes were isolated, however (to be discussed in results), then Lorentzian peaks described the data better.

SPECIFIC EXAMPLES

Changes in reaction conditions such as sorbent phase, pH, surface loading, and ionic strength all have a pronounced effect on sulfate adsorption mechanisms on iron oxides. Therefore, the effects of changing these reaction parameters will be discussed separately.

Influence of pH

On goethite (Figure 8.4a), sulfate adsorption mechanisms are highly pH dependent. As the pH is lowered from pH 9 to pH 6, sulfate adsorption increases, but in all cases only one broad peak (v_3) at approximately 1108 cm^{-1} is present. This peak is consistent with outer-sphere sulfate surface complexes, since outer-sphere sulfate would be expected to retain its T_d symmetry. Also consistent with an outer-sphere complex is the appearance of a v_1 peak at 975 cm^{-1} at pH 6.0 and the systematic shift of the v_3 band to higher wave numbers as pH decreases. This is typical of an electrostatic interaction, because as pH is decreased, the surface charge of the surface becomes more positive, and increased distortion of the T_d symmetry is expected. As the pH is decreased below 6.0, splitting of the 1050 to 1200 cm^{-1} region into multiple peaks occurs, and the peak at 975 cm^{-1} becomes much larger, along with a smaller peak at approximately 1000 cm^{-1}. This is strong evidence of an inner-sphere surface complex. In fact, peak-fitting revealed that there are two sulfate surface complexes at low pH. The inner-sphere complex of C_1 or possibly C_{2v} symmetry has three v_3 bands that appear at 1170, 1134, and 1051 cm^{-1} and a v_1 band occurring at 992 cm^{-1}. The outer-sphere complex also seen at high pH remains in the low pH samples, but the v_3 band continues to shift to higher wave number as pH is lowered. At pH 3.5, this peak occurs at 1118 cm^{-1}, a shift of 14 cm^{-1} from the same 1104 cm^{-1} peak at pH 8.0.

These results suggest that rather than the outer-sphere sulfate converting to an inner-sphere complex at low pH, instead there is additional adsorption of an inner-sphere component. This is clearly demonstrated in Figure 8.4b, where the spectra

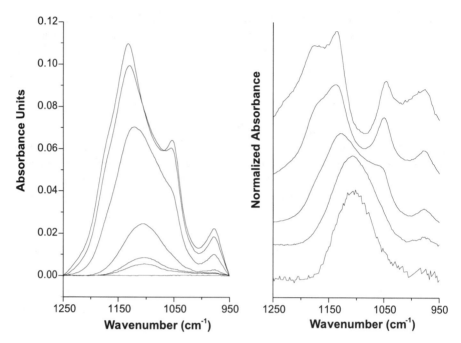

Figure 8.4   (a) Spectra from a pH envelope of sulfate adsorbed on goethite. Reaction conditions were 0.01 N NaCl as background electrolyte, and 20 μM SO$_4^{2-}$ added. The spectra were the result of 128 co-added scans at 4 cm^{-1} resolution. The spectra were collected at pH (from bottom): pH 8, 7, 6, 5, 4, and 3.5 (b) Difference spectra obtained by subtracting pH 8.0, 7.0, 6.0, 5.0, and 4.0 spectra from the pH 7, 6, 5, 4, and 3.5 spectra, respectively, of Figure 8.4a.

from pH 8, 7, 6, 5, and 4 are subtracted from pH 7, 6, 5, 4, and 3.5, respectively. These "difference" spectra are equal to spectra of the additional sulfate adsorption that occurs as pH is lowered. Using this method, it can clearly be seen that above pH 6, all adsorption is due to outer-sphere complexation. However, in the pH 5 minus pH 6 spectrum (c), it is clear that the additional adsorption has a large inner-sphere component. And in the pH 4 minus pH 5 spectrum (b), the additional adsorption can be described only with an inner-sphere complex. Interestingly, in the pH 3.5 minus pH 4 spectrum (a), the same inner-sphere peaks are seen as in (b) but with negative absorbance occurring in the 1050 to 1100 cm^{-1} range. This is due either to desorption of outer-sphere sulfate or to a transformation of some outer-sphere sulfate into an inner-sphere complex. One other important difference that can be seen in Figure 8.4b involves the v_3 peak occurring at the highest wave number. This peak can weakly be seen in (c) at approximately 1160 cm^{-1}, in (b) at 1170 cm^{-1}, and in (a) at 1175 cm^{-1}. The observation that the position of this peak varies with pH would suggest that the peak is the result of an interaction of an electrostatic nature.

The results of fitting two spectra from the pH envelope of Figure 8.4 are shown in Figure 8.5. Figure 8.5a shows the peak fitting results from the pH 3.5 spectrum of Figure 8.4a, and Figure 8.5b shows the results of fitting the pH 5 minus pH 4

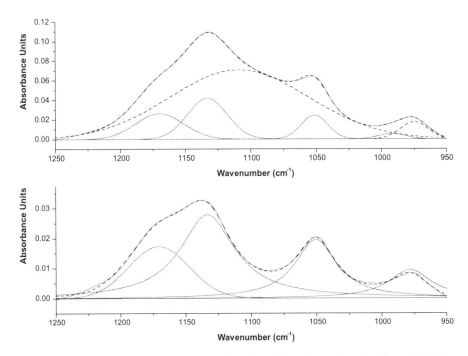

Figure 8.5   (a) Fit of the pH 3.5 spectrum from the pH envelope shown in Figure 8.4a. The dotted lines denote the peaks arising from an inner-sphere complex, and the dashed lines are from the outer-sphere sulfate. (b) Fit of the pH 4 minus pH 5 difference spectrum (b of Figure 8.4b).

spectrum from Figure 8.4b. To accurately fit the spectrum in Figure 8.5a, Gaussian peaks were needed for all peaks. The dotted lines are peaks that are definitely associated with an inner-sphere complex, and the dashed peak centered at 1108 cm^{-1} is due to outer-sphere sulfate. The dashed peak at 976 cm^{-1} is less conclusively assigned, as both outer-sphere and inner-sphere sulfate have ν_1 absorbance bands in this region. It is interesting to fit the "pH 5 minus pH 4 difference" spectrum because there is no contribution from outer-sphere sulfate needed to describe its features. This means that this is a pure inner-sphere sulfate species. While the peaks at 1170, 1132, 1050, and 976 cm^{-1} remain in the same position, the quality of fit was improved by fitting peaks at 1132, 1050, and 976 cm^{-1} with Lorentzian peaks rather than Gaussian peaks. The peak at 1170 cm^{-1} was still best described with a Gaussian peak. This suggests that in the case of the 1170 cm^{-1} feature there is a high degree of disorder, such as an electrostatic bonding environment being responsible for the peak. In the spectrum of Figure 8.5a, the need to describe the inner-sphere complex with Gaussian peaks is likely the result of more than one inner-sphere complex of similar symmetry being present and contributing to the observed peaks. This is supported by the observation that a peak at 992 cm^{-1} in Figure 8.5a is consistent with a ν_1 band resulting from an inner-sphere complex, but it is not present in the difference spectrum of Figure 8.5b. This will be discussed later when the effect of surface loading at low pH is discussed.

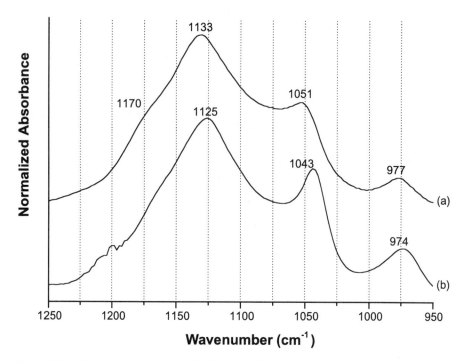

Figure 8.6   Comparison of spectra of adsorbed sulfate collected in (a) H_2O and (b) D_2O. In both cases, the reaction conditions were pH 3.5, I = 0.05, and an initial sulfate concentration of 100 μM.

To further refine the nature of the inner-sphere sulfate surface complex on goethite, sulfate adsorption on goethite at pH 3.5 was studied in D_2O. Using D_2O rather than H_2O can determine the importance of a proton to the surface complex. The results are shown in Figure 8.6. It can clearly be seen that conducting the experiment in D_2O causes the peaks occurring at 1133 and 1051 cm^{-1} in H_2O (splitting of the v_3 band due to inner-sphere complexation) to be shifted approximately 8 cm^{-1} to 1125 and 1043 cm^{-1} in D_2O. This shift to lower wave numbers is characteristic of substitution of a deuterium ion for a proton in a molecular complex. Since these peaks are the result of inner-sphere complexation, it is therefore reasonable to conclude that a proton is present in the inner-sphere surface complex. The only possible surface complexes that involve a proton are bisulfate sorbed as a monodentate complex or a monodentate sulfate that is hydrogen bonded to an adjacent surface site. The fact that the shift is rather small suggests that the complex most likely involves an electrostatic interaction with a proton. This is because direct covalent bonding between a proton and sulfate (such as bisulfate) results in a large splitting in the spectra compared to complexation with Fe (III) in aqueous solution (Hug, 1997). Additionally, the systematic shift of the peak at 1160 to 1175 cm^{-1} in Figure 8.4b to higher wave number as pH is lowered, along with the need to use a Gaussian peak to fit this feature, suggest an electrostatic interaction.

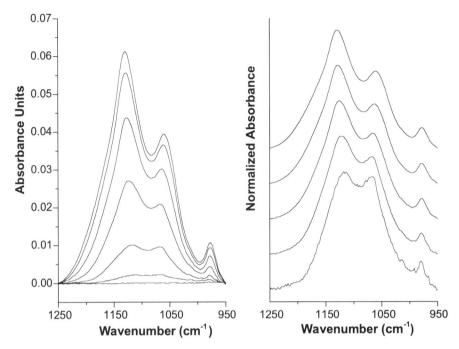

Figure 8.7 (a) Spectra from a pH envelope of sulfate adsorbed on hematite. Reaction conditions were 0.01 N NaCl as background electrolyte, 20 μM SO_4^{2-} added. The spectra were the result of 128 co-added scans at 4 cm^{-1} resolution. The spectra were collected at pH (from bottom): pH 8, 7, 6, 5, 4, and 3.5 (b) Difference spectra obtained by subtracting pH 8.0, 7.0, 6.0, 5.0, and 4.0 spectra from the pH 7, 6, 5, 4, and 3.5 spectra, respectively, of Figure 8.7a.

Influence of Sorbent

On hematite, sulfate adsorption mechanisms are surprisingly different. Figure 8.7a shows spectra from a pH envelope of sulfate adsorption on hematite. Our results are in excellent agreement with those reported by Hug (1997). Strong splitting of the sulfate v_3 band into two peaks at 1126 cm^{-1} and 1060 cm^{-1}, and a clearly visible v_1 peak at 976 cm^{-1} can be seen in all spectra. This C_{3v} symmetry suggests that sulfate forms primarily inner-sphere monodentate surface complexes on hematite over all studied pH values. When difference spectra are produced via the same method discussed for the sulfate/goethite envelope in Figure 8.4b, it can be seen that, generally speaking, a monodentate surface complex indeed dominates the entire pH range studied. However, there appears to be some outer-sphere sulfate adsorption that causes the pH 7 minus pH 8 spectrum (e) to have less splitting of the two v_3 peaks than is seen in the lower pH difference spectra.

Both hematite and goethite are common crystalline iron (III) oxides, but in natural systems amorphous iron hydroxide can also be an extremely important

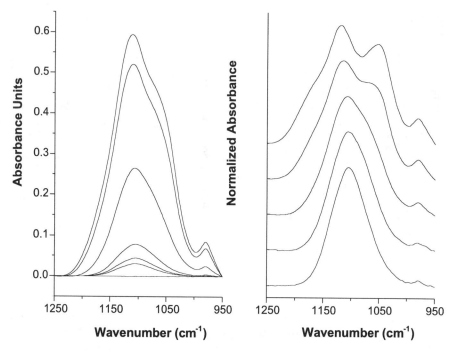

Figure 8.8 (a) Spectra from a pH envelope of sulfate adsorbed on ferrihydrite. Reaction conditions were 0.01 N NaCl as background electrolyte, 20 μM SO_4^{2-} added. The spectra were the result of 128 co-added scans at 4 cm^{-1} resolution. The spectra were collected at pH (from bottom): pH 8, 7, 6, 5, 4, and 3.5 (b) Difference spectra obtained by subtracting pH 8.0, 7.0, 6.0, 5.0, and 4.0 spectra from the pH 7, 6, 5, 4, and 3.5 spectra, respectively, of Figure 8.8a.

sorbent. On 6-line ferrihydrite, sulfate adsorption is substantially different from either goethite or hematite. Figure 8.8a shows a pH envelope of sulfate adsorption on ferrihydrite at an ionic strength of 0.01 M and C_{eq} of 20 μM sulfate. It is clear from the intensity of absorbance that far more sulfate adsorption occurs on ferrihydrite than on either of the crystalline iron oxides. This is to be expected, as the surface area of ferrihydrite is much greater than goethite or hematite. Also readily apparent is the fact that far more outer-sphere sulfate adsorption occurs on ferrihydrite. In fact, the spectra collected at pH 6, 7, and 8 perfectly match the outer-sphere sulfate reference from Nakamoto (1986). As the pH decreases further, more splitting of the v_3 band can be seen. To determine whether this splitting is simply due to the increase of the positive surface charge of the sorbent as pH is lowered or due to the presence of a second inner-sphere complex, the difference spectra were also analyzed (Figure 8.8b). Looking at the pH 3.5 minus pH 4 spectrum and the pH 4 minus pH 5 spectrum (a and b, respectively), it can be seen that between pH 3.5 and 5, the spectra of the additional sulfate adsorption has peaks at 1115 and 1050 cm^{-1} and that a shoulder also appears around 1170 cm^{-1}. Such a large degree of splitting in the v_3 band almost certainly results from an inner-sphere complex. The degree of splitting of the v_3 band is less than when inner-sphere complexes form on goethite,

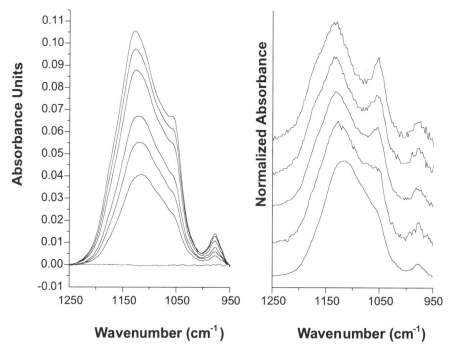

Figure 8.9 (a) Spectra from a sulfate adsorption isotherm on goethite conducted at pH 5.0 and I = 0.01. The spectra are the result of (from bottom): (a) 0, (b) 5, (c) 10, (d) 25, (e) 100, (f) 250, and (g) 500 μM equilibrium sulfate concentration. (b) Difference spectra obtained from Figure 8.9a showing the mechanism of additional sulfate adsorption as loading increases.

but that could possibly be an artifact of the method of taking difference spectra. If additional outer-sphere sulfate also occurs (which appears to be the case on ferrihydrite), then the peaks of the inner-sphere complex will not be very well resolved in the resulting difference spectra.

Influence of Sulfate Concentration

In the case of sulfate adsorption on hematite, Hug (1997) reported that the same adsorption mechanism (an inner-sphere monodentate surface complex) occurs over a wide range of sulfate surface loading. On goethite, however, more than one adsorption mechanism is present. Therefore, shifts from outer-sphere to inner-sphere surface complexes might occur as loading increases. To investigate this, an adsorption isotherm was conducted at pH 5.0 and an ionic strength of 0.01 M (Figure 8.9). At this pH, there is a mixture of outer- and inner-sphere adsorption occurring. When one takes difference spectra from successively higher loading, then it becomes clear that as loading increases the additional adsorption occurs primarily due to an inner-sphere mechanism. At an equilibrium sulfate concentration less than 100 μM, both outer-sphere and inner-sphere complexation increase as loading increases, although

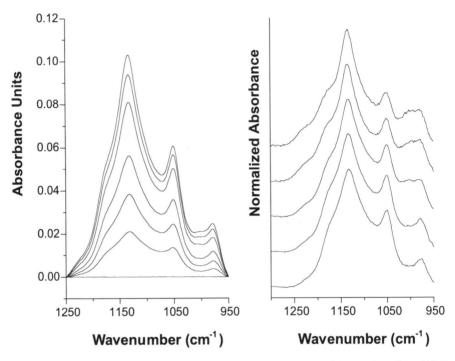

Figure 8.10   (a) Spectra from a sulfate adsorption isotherm on goethite conducted at pH 3.5 and I = 0.05. The spectra are the result of (from bottom): (a) 0, (b) 5, (c) 10, (d) 25, (e) 100, (f) 250, and (g) 500 μM equilibrium sulfate concentration. (b) Difference spectra obtained from Figure 8.10a showing the mechanism of additional sulfate adsorption as loading increases.

the increase in inner-sphere surface complexation is greater than that of outer-sphere complexation. At equilibrium sulfate concentration above 100 μM, essentially all of the additional sulfate adsorption is inner-sphere. This mechanism seems to remain the same as the surface loading increases, because the v_3 inner-sphere peaks at 1050 and 1133 cm^{-1} remain in the same position and the shoulder at 1170 cm^{-1} is very weakly visible in both (a) and (b).

On goethite at pH 3.5, both outer-sphere and inner-sphere surface complexes are seen, with much more inner-sphere adsorption occurring than at pH 5.0. To investigate the potential surface loading effects on inner-sphere sulfate complexation at pH 3.5, an ionic strength of 0.05 M was used to decrease the amount of outer-sphere complexation. The results of this isotherm can be seen in Figure 8.10. If one looks only at the raw spectra (Figure 8.10a), it appears that no significant changes are occurring with increasing surface loading; but when difference spectra are examined (Figure 8.10b), it becomes clear that increasing the sulfate equilibrium concentrations does have an effect on surface complexation mechanisms. Two peaks at 1228 and 1000 cm^{-1} become visible as loading increases. These peaks suggest that another inner-sphere complex is present at higher loading. The fact that the v_1 band is shifted to 1000 cm^{-1} would require the surface complex to have C_{2v} symmetry. There are two potential complexes that satisfy this; bidentate bridging (binuclear) and bidentate chelating sulfate. From references in Figure 8.2, if the complex were the result of a chelating sulfate, then all of the v_3 bands would be expected to shift to higher wave number. While the feature at 1225 cm^{-1} does occur at a relatively high wave number for a sulfate v_3 peak, the other v_3 bands of this complex apparently overlap with those of the inner-sphere complex with C$_1$ symmetry (1050 and 1133 cm^{-1}) since no additional peaks are observed. This suggests that the bidentate binuclear inner-sphere surface complex is responsible for the observed features.

Influence of Ionic Strength

Sulfate adsorption on goethite is influenced by ionic strength as well, with more adsorption occurring as ionic strength is decreased. It is important to separate ionic strength effects from surface loading effects that occur. Figure 8.11 compares sulfate adsorbed on goethite at pH 4.0 and different ionic strengths. Because different amounts of goethite were present in both experiments, the spectra were normalized to a maximum of 1. It is clearly shown that the peaks due to inner-sphere complexation become more pronounced as ionic strength increases. This occurs even though the total amount of adsorption is decreased. This is in contrast to the surface loading effect discussed above that caused increased inner-sphere complexation as loading increased. The earlier assignment of the peaks at 1110 cm^{-1} and 976 cm^{-1} to outer-sphere sulfate is also supported by these results, since the relative importance of these peaks is greater in lower ionic strength spectra (a) where there is more outer-sphere adsorption. The assignment of the peak at 976 cm^{-1} to the v_1 band of outer-sphere sulfate is also supported by the observation that the ratio of the outer-sphere peak areas is 0.04:1.00 in both spectra. Since different amounts of goethite result in differences in total adsorption (and therefore absorbance), it was not possible to produce a difference spectrum to better illustrate the ionic strength dependence.

SIGNIFICANCE OF FINDINGS

In the broadest sense, sulfate adsorption on iron (III) oxides is similar for goethite, hematite, and ferrihydrite. Inner-sphere sulfate adsorption on iron oxides increases as pH is lowered, and outer-sphere adsorption increases as ionic strength is decreased. This adsorption behavior can be readily explained by the concept of ligand exchange. In basic conditions, the singly coordinated surface hydroxyls of metal oxides exist as either Me-O$^-$ or as Me-OH functional groups. The bonds between the oxygen ligands and the metal center tend to be strong, and ligand exchange is less favorable since the hydroxide ligands are difficult to displace and are also present in higher concentrations in solution than is a trace adsorbate such as sulfate. As pH decreases, singly coordinated hydroxyls (Fe-OH) protonate to produce Fe-OH$_2^+$ functional groups. The water attached to the iron (III) is a weak ligand with high lability and can more easily be displaced by a competing ligand such as sulfate, forming an inner-sphere surface complex. To fully understand the reactivity of sulfate with goethite, and to explain differences in sulfate adsorption

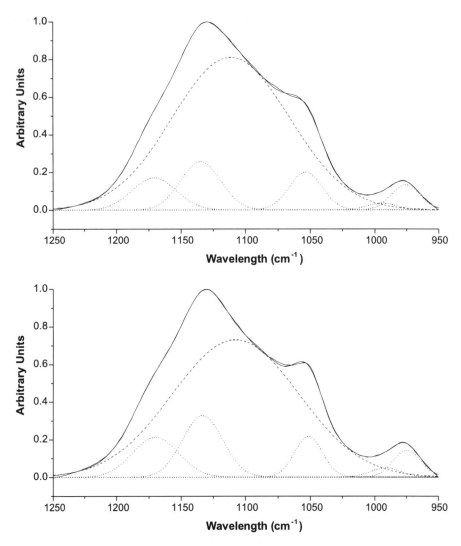

Figure 8.11 Spectra illustrating the variation of sulfate adsorption on goethite with ionic strength. In both cases, the reaction conditions were pH 4.00 and 20 μM SO$_4^{2-}$ added. Spectrum (a) was collected at ionic strength 0.005, while (b) was collected at I = 0.1.

on hematite and goethite, a detailed understanding of the surface-charging behavior of both iron oxides is required.

It is noteworthy that sulfate only forms inner-sphere monodentate surface complexes on hematite in aqueous solution (Hug, 1997) while forming a mixture of outer-sphere and inner-sphere complexes on goethite (Peak, 1999). Since both the goethite used in this experiment and the hematite used by Hug (1997) had a PZC between 8 and 8.5, and in both cases the reactive site is probably Fe-OH$_2^+$, it is somewhat surprising that their reactivities are so markedly different. Differences apparently exist between the populations of functional groups at the metal oxide surface that can account for the different sulfate adsorption behavior observed in the two systems. Researchers have successfully applied a Multi Site Complexation (MUSIC) approach to determine the relative site densities and log K values for all functional groups present on various crystal faces for both goethite and hematite. The results of this analysis can provide valuable insight into the observed adsorption behavior of sulfate, and are used as a point of reference for the discussion that follows.

On goethite, the 110-crystal face is dominant, and the sites that readily protonate are the singly coordinated (Fe-O) and about 67% of the triply coordinated (Fe_3O) surface groups. At the PZC (around 9.0) the singly coordinated surface groups exist as Fe-OH due to the extremely high log K, and the Fe_3O sites are ~70% Fe_3OH. As the pH is lowered from the point of zero charge, only the Fe_3O sites protonate until about pH 7. This explains the observed outer-sphere complexation, because this protonation increases the positive charge on goethite (Fe_3OH has +½ formal charge) and therefore increases sulfate adsorption. However, no $FeOH_2^+$ sites necessary for inner-sphere sulfate complexation appear until the pH is lowered further.

With hematite, the surface functional groups behave somewhat differently. The 110 and the 001 faces predominate. Since the 001 crystal face has a PZC close to that of the hematite used by Hug (1997), this crystal face likely contributed more to the results of Hug's experiments than the 110 face with its PZC of 11. It has also been determined that sulfate forms bidentate binuclear surface complexes on most crystal faces, and monodentate surface complexes only on the 001 planes. On the 001 face, surface functional groups occur due to imperfections in the crystal structure. Both Fe-O and Fe_2O functional groups exist. At the PZC, approximately 80% of the singly coordinated sites exist as $Fe-OH_2^+$ and are capable of ligand substitution reactions with oxyanions. As pH is decreased below the PZC, the remainder of the Fe-OH sites protonate, and the Fe_2O^- sites protonate to form the neutral Fe_2OH. Since the reactive surface sites are formed at a much higher pH on hematite and protonation of the Fe_2O sites does not promote outer-sphere adsorption, inner-sphere surface complexes are more favorable. This can explain Hug's observation that sulfate primarily forms inner-sphere surface complexes over all pH values on hematite. The major points of this discussion are summarized in Figures 8.12 and 8.13.

Unfortunately, no CD-MUSIC proton affinity data exist for ferrihydrite due to its amorphous nature. However, recent research into the structure of ferrihydrite provides a potential explanation for sulfate adsorption on this sorbent phase. Ferrihydrite has an extremely small particle size (roughly 30 angstroms). This means that much more of the total mass of ferrihydrite is reactive surface area than crystalline iron oxides, in which much of the mass is present as a structural and unreactive mineral phase (Zhao et al., 1994). The lack of crystalline order also allows for greater surface site density. These differences between ferrihydrite and both goethite and hematite could be responsible for the observed tendency of sulfate to adsorb to ferrihydrite via an outer-sphere complexation mechanism relative to the crystalline sorbents. It has been discussed that as surface loading increases on goethite, sulfate adsorption mechanisms become increasingly inner-sphere. This can reasonably be

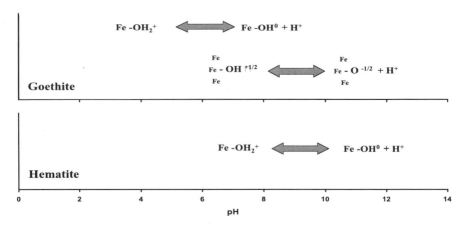

Figure 8.12   Illustration of the CD MUSIC model's description of surface hydroxyl protonation on goethite and hematite as a function of pH.

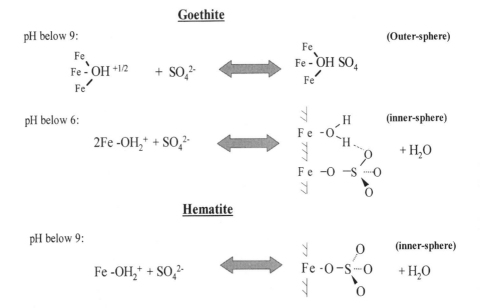

Figure 8.13   Illustration of sulfate surface complexation mechanisms on goethite and hematite as determined via ATR-FTIR spectroscopy.

expected to occur on ferrihydrite as well since a mixture of outer- and inner-sphere surface complexes are similarly seen. It is worth noting that the surface area of ferrihydrite is usually between 200 and 300 m^2g^{-1} (Schwertmann and Cornell, 1991) — roughly 5 times that of the goethite used in this experiment, which had a surface area of 63 m^2g^{-1}. The maximum absorbance of sulfate in the pH envelopes at pH 3.5 is also about 5 times greater (0.6 absorbance units for ferrihydrite vs. 0.11

for goethite). However, this identical loading on a surface area basis could easily result in a much lower degree of surface site saturation if the density of the surface sites of ferrihydrite was significantly greater than goethite. Another possible reason for differences in complexation mechanisms between ferrihydrite and other iron oxides could be related to the structure of the mineral phase. While it is difficult to determine the exact structure of ferrihydrite, several researchers have used EXAFS to probe the structure of ferrihydrite (Zhao et al., 1994). It has been determined that some surface irons are coordinatively unsaturated and, therefore, chemisorb water molecules. If the bonding between these Lewis acid sites and the adsorbed water is strong enough, it is possible that sulfate would simply adsorb as an outer-sphere complex to this positively charged site rather than displacing it via ligand exchange.

Dzombak and Morel (1990) extensively studied the adsorption of cations and anions to ferrihydrite and modeled the reactions using a generalized double-layer model. They concluded that only one site was needed to model proton adsorption based on titration data, but that two sites (a high-affinity site and a low-affinity site) were necessary to describe metal adsorption. They concluded that one site was sufficient to describe anion complexation with ferrihydrite as well, with multiple surface complexes occurring on that single type of site. This theory that sites on the ferrihydrite surface can have different reactivity while having identical pK_a values is also a potential explanation for the observed mixture of outer-sphere and inner-sphere surface complexation of sulfate.

Another point for discussion is the difference in the identity of the inner-sphere surface complexes that form on goethite, hematite, and ferrihydrite. The most reasonable explanation is simply that on the goethite surface the density of surface sites makes it possible for sulfate to form hydrogen bonds with adjacent sites when it adsorbs as an inner-sphere complex. Since this hydrogen bonding would lessen the effective negative charge that a monodentate sulfate surface complex adds to the surface, then it would be favored. It could be that on hematite the appropriate surface configuration simply does not exist in large enough amounts to influence monodentate bonding.

On goethite, the effects that ionic strength and solution sulfate concentration have on sulfate complexation also deserve discussion. These are two distinct effects, because increasing surface loading by increasing the equilibrium sulfate concentration results in an increase in inner-sphere sulfate complexation. However, increasing the surface loading by lowering the ionic strength of the system results in additional outer-sphere complexation. This is because, in the case of ionic strength effects, there is direct competition between sulfate and Cl^- ions that are used as background electrolytes for outer-sphere adsorption on goethite. Additionally, as ionic strength is increased, the electrical double layer is compressed and this also decreases the amount of sulfate that can form outer-sphere complexes. When ionic strength is held constant and loading is increased by raising the concentration of sulfate in solution, then additional adsorption occurs simply to maintain equilibrium between aqueous and adsorbed sulfate. Furthermore, because additional inner-sphere adsorption of the sulfate anion leads to decreased surface charge, inner-sphere complexation is favored. At pH 3.5, there is an additional effect of surface loading: the formation of an additional inner-sphere surface complex of C_{2v} symmetry. This complex seems most likely to be a bidentate bridging complex, and probably forms as a result of surface crowding. A bidentate surface complex takes up less area on the surface than a monodentate complex even if the monodentate complex is hydrogen bonded to an adjacent surface site. Alternatively, this bidentate bridging surface complex could be the result of adsorption to less favorable surface sites that only occurs when a large solution sulfate concentration drives the reaction.

It is also noteworthy that on both goethite and ferrihydrite, lowering solution pH generally results in inner-sphere surface complexation rather than a transformation from an outer-sphere surface complex to an inner-sphere surface species. The only exception that was seen in this research was when pH was lowered from 4 to 3.5 on goethite (Figure 8.4). Under this condition, a negative absorbance in the region where outer-sphere sulfate occurs was noted. This possibly could be due to some transformation to an inner-sphere sulfate on the surface. Alternatively, this could be explained by desorption of outer-sphere sulfate caused by decreased surface charge on the surface.

OVERALL CONCLUSIONS

Sulfate adsorption on iron oxides and hydroxides is quite complex. It is important to consider not only the effects of pH, ionic strength, and reactant concentration on sulfate adsorption, but also the nature of the sorbent phase being studied. Sulfate forms inner-sphere monodentate surface complexes on hematite from pH 8 to 3.5 and across a wide range of surface loadings. On goethite, however, sulfate forms only outer-sphere surface complexes at pH 6 and above, and forms a mixture of outer-sphere and inner-sphere complexes below pH 6. The inner-sphere sulfate surface complex is a monodentate complex that is hydrogen bonded to an adjacent surface site. Increasing the equilibrium sulfate concentration promotes additional inner-sphere sulfate adsorption at pH 4.5 to 6. At pH 3.5, raising the sulfate equilibrium concentration causes the formation of a second bidentate binuclear surface complex. Decreases in ionic strength lead to increased sulfate adsorption, with the additional sulfate being primarily outer-sphere. Finally, sulfate forms predominantly outer-sphere surface complexes on ferrihydrite. Some inner-sphere complexation occurs below pH 5.0, with the spectra suggesting a monodentate surface complex. The amount of adsorption on ferrihydrite is much greater than either of the crystalline iron oxides in this study. Much research in the scientific literature involves experiments conducted with goethite and hematite that was not treated to remove amorphous iron oxides prior to adsorption studies. Since ferrihydrite was found to have far more outer-sphere adsorption of sulfate than either goethite or sulfate, this could potentially result in incorrect assessment of adsorption mechanisms on these sorbents.

There are also some more important conclusions that can be drawn from this study. It is reasonable to assume that, similar to sulfate, other oxyanions may also have very different adsorption mechanisms on hematite, goethite, and ferrihydrite. It is, therefore, of great importance to study how changes in surface functional group distributions and log Ks of iron oxides affect complexation mechanisms. Another important concept that seems to be becoming clearer to soil chemists is that in many systems multiple adsorption mechanisms are occurring simultaneously and that a continuum of adsorption processes exists. This is not only true for sulfate; it has been shown that organic ligands (Persson et al., 1998) can adsorb as both outer-sphere or inner-sphere complexes as reaction conditions change. Metal ions also exhibit this trend, as competition between adsorption and precipitation has been seen in the case of mixed metal precipitate formation (Elzinga, 2000) on some clay minerals. It therefore makes sense to use molecular-scale spectroscopy to study the effects that changing reaction conditions have on this continuum of reaction mechanisms.

REFERENCES

Ali, M. A. and D. A. Dzombak (1996). Competitive sorption of simple organic acids and sulfate on goethite. *Environ. Sci. Technol.* 30(4): 1061-1071.

Charlet, L., N. Dise, and W. Stumm (1993). Sulfate adsorption on a variable charge soil and on reference minerals. *Agriculture, Ecosystems, and Environment.* 47: 87-102.

Davis, J. A. and J. O. Leckie (1980). Surface ionization and complexation at the oxide/water interface. 3. Adsorption of anions. *J. Colloid Interface Sci.* 74: 32-43.

Degenhardt, J. and A. J. McQuillan (1999). *In situ* ATR-FTIR spectroscopic study of adsorption of perchlorate, sulfate, and thiosulfate ions onto chromium(III) oxide hydroxide thin films. *Langmuir.* 15: 4595-4602.

Dzombak, D. A. and F. M. M. Morel. (1990). *Surface Complexation Modeling: Hydrous Ferric Oxide.* John Wiley & Sons. New York.

Eggleston, C. M., S. Hug, W. Stumm, B. Sulzberger, M. Dos Santos Alfonso (1998). Surface complexation of sulfate by hematite surfaces: FTIR and STM observations. *Geochim. Cosmochim. Acta.* 62(4): 585-593.

Elzinga, E. J. and D. L. Sparks (2000). Reaction condition effects on nickel sorption mechanisms at the illite/water interface. *Soil Sci. Soc. Am. J.* In press.

Geelhoed, J. S., T. Hiemstra, W. H. van Riemsdijk (1997). Phosphate and sulfate adsorption on goethite: single anion and competitive adsorption. *Geochim. Cosmochim. Acta.* 61(12): 2389-2396.

He, L. M., L. W. Zelazny, V. C. Baligar, K. D. Ritchey, and D. C. Martens. (1996). Hydroxyl-sulfate exchange stoichiometry on γ-Al_2O_3 and kaolinite. *Soil Sci. Soc. Am. J.* 60: 442-452.

He, L. M., L. W. Zelazny, V. C. Baligar, K. D. Ritchey, and D. C. Martens (1997). Ionic strength effects on sulfate and phosphate adsorption on γ-alumina and kaolinite: triple layer model. *Soil Sci. Soc. Am. J.* 61: 784-793.

Hodges, S. C. and G. C. Johnson (1987). Kinetics of sulfate adsorption and desorption by Cecil soil using miscible displacement. *Soil Sci. Soc. Am. J.* 51: 323-331.

Hug, S. J. (1997). *In situ* Fourier transform infrared measurements of sulfate adsorption on hematite in aqueous solutions. *J. Colloid Interface Sci.* 188: 415-422.

Martin, R. R. and R. S. C. Smart. (1987) XPS photoelectron studies of anion adsorption on goethite. *Soil Sci. Soc. Am. J.* 51: 54-56.

Nakamoto, K. (1986). *Infrared and Raman Spectra of Inorganic and Coordination Compounds.* John Wiley & Sons, New York.

Parfitt, R. L. and R. S. C. Smart (1978). Mechanism of sulfate adsorption on iron oxides. *Soil Sci. Soc. Am. J.* 42(1): 48-50.

Peak, D., R. G. Ford, and D. L. Sparks. (1999). An *in situ* ATR-FTR investigation of sulfate bonding mechanisms on goethite. *J. Colloid Interface Sci.* 218: 289-299.

Persson, P., P. Nordin, J. Rosenqvist, L. Lövgren, L. Öhman, and S. Sjöberg (1998). Comparison of the adsorption of *o*-phthalate on boehmite, aged γ-Al_2O_3, and goethite. *J. Colloid Interface Sci.* 206: 252-266.

Persson, P. and L. Lövgren (1996). Potentiometric and spectroscopic studies of sulfate complexation at the goethite-water interface. *Geochim. Cosmochim. Acta.* 60(15): 2789-2800.

Ridley, M. K., D. J. Wesolowski, D. A. Palmer, P. Benezeth, and R. M. Kettler (1997). Effect of sulfate on the release rate of Al^{3+} from gibbsite in low-temperature acidic waters. *Environ. Sci. Technol.* 31: 1922-1925.

Rietra, R. P. J. J., T. Hiemstra, W. H. van Riemsdijk (1999). Sulfate adsorption on goethite. *J. Colloid Interface Sci.* 218: 511-521.

Schwertmann, U., P. Cambier, and E. Murad (1985). Properties of goethites of varying crystallinity. *Clays Clay Miner.* 33(5): 369-378.

Schwertmann, U. and R. M. Cornell (1991). *Iron Oxides in the Laboratory: Preparation and Characterization.* New York, NY, Weinheim.

Sparks, D. L. (1995). *Environmental Soil Chemistry.* San Diego CA, Academic Press. 1995.

Sposito, G. A. (1984) *The Surface Chemistry of Soils.* New York, Oxford University Press.

Stumm, W. and J. J. Morgan (1996). *Aquatic Chemistry.* New York, John Wiley & Sons. 1996.

Sugimoto, T. and Y. Wang (1998). Mechanism of the shape and structure control of monodispersed α-Fe2O3 particles by sulfate ions. *J. Colloid Interface Sci.* 207(1): 137-149.

Turner, I. J. and J. R. Kramer (1991). Sulfate ion bonding on goethite and hematite. *Soil Sci.* 152(3): 227-230.

Venema, P., T. Hiemstra, Peter Weidler, and W. H. van Riemsdijk (1998). Intrinsic proton affinity of reactive surface groups of metal (hydr)oxides: application to iron (hydr)oxides. *J. Colloid Interface Sci.* 198(2): 282-295.

Yamaguchi, N. U., M. Okazaki, and T. Hashitani (1999). Volume changes due to SO_4^{2-}, SeO_4^{2-}, and $H_2PO_4^-$ adsorption on amorphous iron (III) hydroxide in an aqueous suspension. *J. Colloid Interface Sci.* 209: 386-391.

Yates, D. E. and T. N. Healy (1975). Mechanism of anion adsorption at the ferric and chromic oxide/water interfaces. *J. Colloid Interface Sci.* 52: 222-228.

Zhang, P. C. and D. L. Sparks (1990). Kinetics and mechanisms of sulfate adsorption/desorption on goethite using pressure-jump relaxation. *Soil Sci. Soc. Am. J.* 54: 1266-1273.

Zhao, J. M., F. E. Huggins, Z. Feng, G. P. Huffman (1994). Ferrihydrite — surface-structure and its effect on phase transformation. *Clays Clay Miner.* 42: (6) 737-746.

CHAPTER 9

Selenium Contamination in Soil: Sorption and Desorption Processes

B. Pezzarossa and G. Petruzzelli

BIOLOGICAL IMPORTANCE OF SELENIUM

The biological importance of selenium is mainly linked to three factors: it is an essential element in animal and, probably, vegetable metabolisms; in many geographical areas, the available quantity is insufficient to satisfy animal requirements; in some areas, it is present in such high concentrations in soil, water, plant, ash, and aerosol that it is toxic for animals.

In humans and animals, selenium can be either beneficial, in some cases essential (Underwood, 1977), or toxic (Yang et al., 1983), depending on its concentration. Its importance in human nutrition is now accepted. Selenium, which acts as a metal co-factor of the enzyme glutathione peroxidase, inducing the reduction of lipid hydroperoxides and hydrogen peroxide, has been identified in human serum, urine, blood, and scalp hair. Moreover, there is experimental evidence for its anticarcinogenicity (Chortyk et al., 1984) and for its effects in the neutralization of the toxicity of heavy metals (Vokal-Borek, 1979). In order to prevent Se-deficiency, which reduces growth, productivity, and reproduction, dietary intake should be in the range of 0.05 to 0.1 mg Se kg^{-1}, but Se toxicity may appear when dietary levels exceed 5 to 15 mg Se kg^{-1}. Major sources of Se toxicity are inhalation or dermal contact, uncontrolled self-medication, and high levels of dietary intake mostly associated with people farming over seleniferous soils. Loss of hair and nails, following nausea, and diarrhea are common symptoms of Se toxicity (Mayland, 1994).

Plants do not require Se, but absorb it from soil solution and recycle it to ingesting animals. Selenium is taken up by plants and incorporated into amino acids and proteins (Shrift, 1973). The levels of accumulation in plants depend on the amount of available selenium, the pH value, salinity and $CaCO_3$ content of the soil, and on plant species.

1-56670-531-2/01/$0.00+$1.50
© 2001 by CRC Press LLC

Table 9.1   Selenium Concentration in Different Materials

Material	Se (mg kg⁻¹)
Earth's crust	0.05
Granite	0.01–0.05
Limestone	0.08
Shales	0.6
Phosphate rocks	1–300
Seleniferous soils	1–80 (up to 1200)
Other soils	0.01–4.7
Coal	0.46–10.65
Atmospheric dust	0.05–10
Rivers	
Mississippi	0.00014
Colorado	0.01–0.4
Plants	
Graminacee	0.01–0.04
Clover	0.03–0.88
Barley	0.2–1.8
Animal tissue	0.4–4

From McNeal, J.M. and L.S. Balistrieri, 1989. Geochemistry and occurrence of selenium: an overview. In *Selenium in Agriculture and the Environment*, Jacobs, L.W., Ed., SSSA Special Publication.

SELENIUM IN THE ENVIRONMENT

The total concentration of selenium, which is found in nearly all materials of the earth's crust, has been determined in rocks, soil, fossils, volcanic gases, waters, and plant and animal tissue (Table 9.1). Biological activity plays an important role in the distribution of selenium in the environment. The accumulation in plants and animals varies enormously and can positively or negatively affect their growth, development, and reproduction.

Selenium is involved in many different physical, chemical and biological processes, including: volcanic activity; combustion of fossil fuels (coal, oil); processing of nonferrous metals; incineration of municipal waste; erosion, and leaching of rocks and soils; groundwater transport; plant and animal uptake and release; sorption and desorption; chemical and biological redox reactions; and the formation of minerals.

Coal and organic-rich sediments tend to have high selenium concentrations, presumably due to Se sorption or complexation by organic matter. The most important source of Se is represented by the weathering of rocks such as shales, which are rich in selenium (0.6 mg kg⁻¹). Selenium can also be found in phosphate rocks, which holds implications for the agricultural environments where phosphate fertilizers are used (Carter et al., 1972).

Natural waters usually contain low levels of Se (< 0.01 mg L⁻¹), except for alkaline waters or waters that leach and drain seleniferous rocks and soils. In the San Joaquin Valley (California, USA) the water draining irrigated land contains up to 4.2 mg Se L⁻¹ as SeO_4^{2-} (Sylvester, 1986). These waters are subsequently carried to the Natural Kesterson Reservoir, where they induce problems of toxicity to the flora and the protected fauna.

Selenium is used in several industrial processes, especially in the electronic and photoelectric industries. Since a variation in light intensity produces a variation in the electric current in selenium, it is also used in the production of photocopiers. It is used in the glass industry to avoid glass coloration by iron and in the rubber industry to increase the resistance to heat and the speed of vulcanization.

Selenium has chemical properties, which are intermediate between those of metals and nonmetals. It has an atomic number of 34 and is located in the oxygen group of the Periodic Table between nonmetallic sulfur and metallic tellurium. Selenium can exist as selenide (Se^{2-}), elemental selenium (Se^0), selenite (SeO_3^{2-}), and selenate (SeO_4^{2-}). Selenide exists in acid and reducing environments, rich in organic matter.

Elemental selenium is stable in reducing environments and can be oxidized to SeO_3^{2-} and to SeO_4^{2-} by microorganisms (Sarathchandra et al., 1981). Since their salts are insoluble and resistant to oxidation, these forms are poorly available for plants and animals.

Selenite is found in mildly oxidizing environments. H_2SeO_3 is a weak acid, and its salts are less soluble than selenates. It is sorbed by iron oxides, amorphous hydroxides and Al sesquioxides, and it can be reduced to elemental selenium by reducing agents or microorganisms which limit its mobility and bioavailability.

Selenate is stable in alkaline and well-oxidized environments. H_2SeO_4 is a strong acid and forms very soluble salts. SeO_4^{2-}, which is the form most easily absorbed by plants, is not sorbed by soil as well as SeO_3^{2-} and is easily leached and transported in groundwaters. The oxidation from SeO_3^{2-} to SeO_4^{2-} in alkaline and well-oxidized environments may increase the selenium mobility and assimilation by plants, albeit slowly.

Selenium has chemical properties that resemble those of sulfur. Se and S are in the same group in the Periodic Table and can exist in the same oxidation states. They can form similar allotropes (monoclinic and rhombic) and similar compounds, especially organic ones. Since they have the same ionic radius (1.98 Å), selenium can substitute sulfur in many inorganic and organic compounds. Due to its several oxidation states, selenium can behave both like an electron donor and an electron acceptor, and this makes it suitable for biologically active systems. Even though Se and S are often geologically exchangeable, in the soil surface they are involved in different geochemical processes. Their different boiling and fusion points and redox potential make it possible to separate Se and S in the environment (Lakin, 1973). Sulfur can be easily oxidized to sulfate, which is very mobile in soils and groundwater, whereas selenium needs stronger oxidizing conditions to turn it into selenate, which is the most soluble form.

SELENIUM IN PLANTS

On the basis of their capacity both to tolerate high levels of selenium and to accumulate unusually high concentrations, plants grown on seleniferous soils can be divided into three groups: 1) Se accumulator or indicator plants, 2) secondary Se absorber plants, and 3) nonaccumulator plants (Rosenfeld and Beath, 1964).

Accumulator plants require selenium for their growth and include many species from *Astragalus, Stanleya, Machaeranthera,* and *Haploppapus,* which can accumulate 100 to 10,000 mg Se kg^{-1}. Secondary Se absorber plants belong to some species of *Astragalus, Atriplex, Gutierrezia,* and *Machaeranthera,* and rarely absorb more than 50 to 100 mg Se kg^{-1}. Nonaccumulator plants, which include some grains, grasses, and most of the cultivated species, do not accumulate more than 50 mg Se kg^{-1} and usually contain 0.05 to 1 mg Se kg^{-1}. Shrift (1973) hypothesized that selenium is an essential microelement in accumulator plants based on the following evidence: accumulator plants grow only on seleniferous soils and accumulate higher quantities of Se than nonaccumulator plants; the growth of accumulator plants is stimulated by adding small amounts of Se to the growth solution, whereas the growth of nonaccumulator plants is inhibited; the assimilation path of Se in the accumulator plants differs substantially from nonaccumulator plants.

In nonaccumulator plants, Se is found in the form of protein-bound selenomethionine, whereas in accumulator plants it is found in a water-soluble and nonprotein form such as Se-methylselenocysteine. It has been hypothesized that Se accumulator species evolved a detoxification mechanism which excluded Se from protein incorporation (Lewis, 1976). In nonaccumulator plants, devoid of this mechanism, Se is incorporated into proteins, resulting in an alteration or inactivation of the protein structure and possible poisoning of plants. Se concentration levels in soils where accumulator plants grow are generally lower than in soils where nonaccumulator plants grow, suggesting that the former absorb more Se. The cultivation of accumulator plants could represent a valid method to remove Se from contaminated lands. Studies conducted on plants of genus *Brassica, Atriplex,* and *Astragalus,* both under protected cultivation and in open fields, highlighted an accumulation of selenium in the plants and a subsequent reduction in Se concentration in the soil (Banuelos et al., 1990).

The uptake and the metabolism of Se in plants are affected by several factors, such as other ions (Cl^-, SO_4^{2-}, PO_4^{3-}), salinity (Gupta et al., 1982), and trace elements. The interactions between selenium and other ions may be due to chemical reactions either in the soil or in the plant, or to the dilution effect due to an increased plant growth.

A moderate concentration of NaCl (1 to 10 mM) in the growth medium can reduce the Se accumulation and growth inhibition of plants.

The relative plant availability of selenate vs. selenite depends on the concentration of competing ions, specifically sulfate and phosphate. The inhibition in the uptake of selenate by sulfate has been studied (Mikkelsen et al., 1989; Bell et al., 1992; Pezzarossa et al., 1999). Studies conducted on perennial ryegrass and strawberry clover (Hopper and Parker, 1999) showed that inhibition of SeO_4^{2-} uptake by SO_4^{2-} was much stronger than that of SeO_3^{2-} by PO_4^{3-}, since the two latter ions are less similar chemically.

The addition of phosphorus to P-deficient soils induces an increase of Se content in the cultivated plants (Carter et al., 1972). Two possible explanations have been given for this phenomenon: 1) P and Se compete with the same fixation sites, and P might substitute Se making it available for plant uptake; 2) the increased amount of selenium taken up by plants could be related to an enhanced root growth, and consequently to a higher soil volume explored, in response to phosphate fertilizations. Moreover, P fertilizers might contain high amounts of selenium, which can be available for plant uptake.

Sulfate has an antagonistic effect on selenium uptake and can reduce its phytotoxicity. Increases in sulfate concentration reduce selenium accumulation both in roots and leaves, but Se translocation from root to shoot appears to be more negatively affected by high sulfate concentration than Se uptake by roots (Pezzarossa et al., 1997; Pezzarossa et al., 1999). The reduced chemical and physical differences between Se and S result in significant biological differences in the plant. The toxic effects of Se in plants, in fact, are mainly due to the uptake and translocation of Se^{4+} and Se^{6+} throughout the plant and to their incorporation into organic constituents. These compounds act as S analogues and interfere with essential biochemical reactions. The absorption of SeO_4^{2-} by roots follows the same transport path as SO_4^{2-}: the two ions compete for the same binding sites within the root cell (Atkins al., 1988). The competition between Se and S is in relation to their concentration in the growth media. If the sulfate levels are low, there could be more of a synergistic effect than a competitive effect, whereas the sulfur content in the leaves increases by increasing the sulfate concentration (Pezzarossa et al., 1997). When the sulfate concentration is low, selenium tends to accumulate in the roots, whereas a higher amount of Se is translocated to the leaves when the sulfate content increases. Se within the plant is metabolized by the enzymes of the sulfur assimilation path, since it has the ability to resemble S. The first step of Se incorporation into an organic compound is the reduction of Se^{6+} to Se^{4+}, a process that occurs mainly in the leaves. Selenite is then incorporated into biomolecules (selenoetheramino acids as Se-methylselenocysteine or Se-methylselenomethionine), which act as Se-analogues of essential S compounds. The Se-amino acids can disturb the normal biochemical reactions and the enzymatic functions of the cell (Mikkelsen et al., 1989). The growth inhibition caused by selenate can be overcome by the addition of sulfate, providing further evidence that selenium toxicity is related to the competitive interactions between S compounds and their Se-analogues.

SELENIUM IN SOIL

Selenium concentration in most soils varies from 0.01 to 2 mg Se kg^{-1}, while in seleniferous areas it has been found to be as high as 1200 mg Se kg^{-1} total selenium and 38 mg kg^{-1} soluble selenium. Where cases of toxicosis of the livestock occur, soil has been found to contain from 1 to more than 10 mg Se kg^{-1}. In some soils of the Hawaiian Islands, levels can be found up to 20 mg Se kg^{-1}, but not available for plant uptake because complexed by minerals of Fe and Al. Selenium biogeochemistry is, in fact, largely controlled by that of Fe, with which it is tightly associated both in the oxidizing and reducing environments. Soils coming from sedimentary rocks generally have a higher Se content than those deriving from igneous rocks.

In cultivated soils 5 mg Se kg^{-1} are considered to be too high, and 0.03 mg Se kg^{-1} too low for optimal crop production.

As for other elements, the selenium concentration in plants does not necessarily correspond to the total content in soil. The chemical forms of Se in soil are selenide (Se^{2-}), elemental selenium (Se^0), selenite (SeO_3^{2-}), selenate (SeO_4^{2-}), and organic selenium. Among these forms, which are strictly correlated to the pH value and the potential oxide-reduction of the soil, the most available form for plants is considered to be the water-soluble fraction. Studies conducted by Jayaweera and Biggar (1996) showed that changes of Eh and pH can induce several transformations of Se in soil and affect Se release into drainage and groundwater systems. During soil reduction, the total soluble Se and selenate decreased, whereas selenite first increased and then decreased. During soil oxidation the total soluble Se and selenate increased, while selenite first increased and then decreased.

The speciation and the form of selenium can change as a result of biological and physical processes: SeO_3^{2-} and SeO_4^{2-} can be reduced to Se amorphous from bacteria and yeasts, which limits its availability for plant uptake.

Soil plays an important role in the cycle of Se in the geoecosystem, since it has the ability to retain Se, avoiding its loss by leaching. Leaching of the soil profile may result in the mobilization of significant quantities of Se, which in turn may achieve hazardous concentrations in surface, drainage, and groundwater (Neal and Sposito, 1989). The physical, chemical, and biological characteristics of soil affect the mobility and availability of different Se forms. Organic matter, $CaCO_3$ content, pH, cation exchange capacity, and Fe oxide minerals affect the sorption of selenium (Singh et al., 1981; Neal et al., 1987).

Inorganic compounds of Se, as with those of other trace elements, including arsenic, mercury and lead, can be biomethylated to volatile compounds such as dimethyl selenide, DMSe, or dimethyldiselenide, DMDSe. Volatilization of Se represents a system which is able to remove Se from the soil and depends on microbial activity, temperature, moisture, and water-soluble Se. Both microorganisms (bacteria, fungi, and yeasts) and plants (Shrift, 1973) can reduce selenite and selenate to volatile species. From autoclaved or sterilized soil, where microbial activity is removed, no volatilization takes place, confirming that volatilization is a biological process. The formation of methylated compounds from animals seems to represent a mechanism of detoxification: the toxicity of the dimethyl selenide is in fact around $\frac{1}{500}$ to $\frac{1}{1000}$ of selenide toxicity. Organic forms such as selenomethionine, whose uptake is under metabolic control, are important sources of available selenium for plants (Abrahms et al., 1990). In some soils, up to 50% of total selenium can be in organic form.

Se sorption, either as selenite or selenate, has been described by the Langmuir curve (Singh et al., 1981; Tan et al., 1994). SeO_4^{2-} is generally less sorbed than SeO_3^{2-} and soils sorb variable amounts of these anions in the order: organic soil>calcareous soil>normal soil>saline soil>alkali soil. Further sorption studies (Fio et al., 1991) carried out in soils with different irrigation and drainage systems, described selenite sorption with the Freundlich equation and indicated that selenate is not sorbed into soil, whereas selenite is rapidly sorbed.

Selenium is mainly associated with iron and manganese oxides and hydroxides, carbonates, and organic matter. The hydroxides in the soil are not in an organized structure, are interdispersed with clay minerals, and can be found to be precipitated covering the soil particles or filling the pores. Soils of the Mediterranean area contain higher amounts of iron oxides, but they are also rich in aluminum and manganese oxides. The specific characteristic of Fe and Mn oxides is an electric charge which varies in relation to the soil pH value, making them able to sorb Se anions according to the net charge. In alkaline conditions, the charge is negative, whereas in acid conditions it is positive.

Studies conducted on the interactions between Se and oxides show that iron selenite is first sorbed by soil, and then selenate, as a coprecipitate, is formed (Lakin, 1973). Afterwards, the theoretical solubility reactions showed that iron selenite and selenate are too soluble or insufficiently stable to exist in most soils (El Rashidi et al., 1987). In alkaline soils, the interactions between selenite and iron oxides are particularly important in controlling the solubility of selenites (Neal, 1995).

Recently it has been found that iron oxides (goethite and emathite) might sorb both selenite and selenate giving "inner-sphere" and "outer-sphere" complexes, respectively. The inner-sphere complexes are formed when a ligand in solution exchanges with a hydroxylic surface group leading to a specific sorption process. At the beginning, the surface is protonated by a proton deriving from the same diprotic acid, which is also the ligand source. Therefore, the H_2O molecule is exchanged by selenite, according to the formula:

$$Sf\ OH + SeO_3^{2-} + H^+ \Rightarrow Sf\ SeO_3^{2-} + H_2O$$

where Sf is the soil surface.

The bond in a inner-sphere complex can be ionic, covalent, or a combination of the two. When the ligand sorption increases, a consequent increase in the amount of hydroxides released by surface sites must be expected, as shown in the case of absorption of selenites by allophane (Rajan and Watkinson, 1979). The formation of inner-sphere complexes has been confirmed by X-rays studies. The sorption complexes are rather strong, since variations of the ionic strength do not affect selenite sorption on iron oxides such as goethite (Hayes et al., 1987).

The outer-sphere complexes form when the H_2O molecule is held between the surface site and the sorbed ligand. These reactions of ion-pair formation can be described as a nonspecific sorption process. In this case, the H_2O molecule is incorporated inside the complex, according to the formula:

$$Sf\ OH + SeO_4^{2-} + H^+ \Rightarrow SeO_4^{2-} + Sf\ H_2O$$

where Sf is the soil surface.

Since the bond is usually electrostatic and much weaker than the ionic or covalent bonds of the inner-sphere complexes, the sorption complex is less stable. Furthermore, the selenate sorption dramatically decreases when the ionic strength of the solution increases. The formation of outer-sphere complexes has been confirmed by X-rays studies (Hayes et al., 1987).

Several studies (Balistrieri and Chao, 1987; Saeki et al., 1995) have suggested that selenite is sorbed more than selenate and in a wider range of pH. Selenate sorption is a result of electrostatic attractions, which are more affected by pH than inner-sphere sorptions typical of selenite. Fe and Al oxides have surfaces with a variable charge in relation to pH: positive to low pH and negative to high pH. This implies a higher sorption process of selenite and selenate at low pH, since both are present as negative ions in the soil solution.

The pH is not the only factor affecting sorption. Temperature, Eh, selenium concentration, and the effect of ions competing for the same sorption sites play an important part in determining the amount of Se sorbed. The redox conditions, together with pH, control the chemical species of Se in the soil. Selenate, for example, is present when pH ranges from 3 to 10, and Eh <1 volts (Mayland et al., 1989). An increase of 10°C in temperature decreases the selenite sorption, as reported by Balistrieri and Chao (1990).

The degree of competition among ions depends on the affinity of the competing anions for the sorbing surfaces as well as the concentration and the nature of the bonds. Phosphate, sulfate, and arsenate are the most common competitive anions, but other compounds, such as electrolytes (NaCl, $CaCl_2$) can compete all the same. A variation in the electrolyte ionic strength can modify the charge both of the sorbing surface and of the sorbing ions. In theory, selenite in inner-sphere complexes should not be affected by the ionic strength, unlike selenate, which is more sensitive to variations in medium ionic strength. There is evidence of hysteresis phenomena in the desorption processes.

The displacement of selenite by phosphate can have agronomic implications, namely, that selenite might be mobilized following phosphate fertilization.

Studies conducted on sorption and desorption processes give conflicting results (Hingston et al., 1972). As an example, selenite sorbed on goethite was not displaceable by NaCl 1 M, unlike selenite sorption on gibbsite, which was completely reversible when NaCl 0.1 M was added.

It has been found that goethite and gibbsite have sites on which both selenite and phosphate can be sorbed and sites on which only one of the two can be sorbed. The presence of specific sites for selenite has been confirmed by an increase in selenium sorption by goethite in the presence of an excess of phosphate ions (Glasauer et al., 1995). However, these results were obtained in a closed system.

Mn oxides have a zero charge point (ZPC), lower than iron oxides, and consequently it is more difficult to bear positive charges on such surfaces compared to iron oxides, in order to attract and hold selenite and selenate. In fact, a low selenite sorption and no selenate sorption have even been recorded (Balistrieri and Chao, 1990).

Selenium sorption processes are influenced by soil carbonates. Calcite is the only mineral carbonate that has been studied in relation to selenium sorption in soil and sediments. Since the ionic radius of Se is too large to replace Ca in the lattice, Se binds with calcite only by the sorption process. Data obtained in experiments conducted in soils rich in calcite give contrasting results. With excess calcite, the selenium sorption increases. This effect may be due to the formation of a surface precipitate, $CaSeO_3 \cdot 2H_2O$, or to an earlier sorption of Ca^{2+} on the surface of soil together with a subsequent increase in positive charges, which then create a higher attraction for Se (Neal et al., 1987).

The Se sorption on carbonates is also affected by pH and by the presence of competitive anions. In the literature, the contrasting results recorded may be due to the different characteristics of the soil, to the different experimental conditions, and to analytical difficulties in determining low concentrations of selenium. Experimental studies show that sorption increases as pH increases (Goldberg and Glaubig, 1988), but there is some evidence to suggest that Se sorption can decrease by increasing the pH (Cowan et al., 1990).

Due to the negative charge of clay, Se is not easily sorbed and because of its ionic radius (1.84Å) it cannot replace smaller ions (Fe, Al) in the lattice. Thus, it can only be sorbed on the edges of clay minerals, in particular kaolinite (Bar-Yosef, 1987). Its broken edges sorb both H^+ and OH^- ions, developing surfaces with variable charges similar to hydroxides, and therefore depending on the pH. Pure clays are not able to sorb the same amounts of Se as oxides and hydroxides. Oxides have a sorption capacity 10 to 30 times higher than clays. The sorption capacity can be modified or increased by the presence of Fe and Mn oxides, and the organic matter on the clay surface. The effect of pH is important: at low pH values, the sorption capacity increases, whereas at high pH values, desorption processes are more likely. Competitive anions, such as phosphate, affect the desorption process of Se, especially when the concentration of phosphate is much higher than selenate (Cowan et al., 1990).

Selenium can be included in humic substances or can be complexed by means of sorption reactions, but it is not yet clear what the mechanisms of binding are. Organic matter plays a role of primary importance in the chemistry of selenium in soil. The content of selenium in soil has been correlated with the content of organic soil matter (Singh et al., 1981; Johnsson, 1989; Gustafsson and Johnsson, 1992). Selenium complexed by soil organic matter is the predominant chemical form in the podzol.

Se sorption can be facilitated by organic matter covering clay particles and by Fe and Mn hydroxides.

SELENIUM SORPTION IN MEDITERRANEAN SOILS

We conducted experiments to study the sorption and the desorption processes of selenate in various soils typical of the Mediterranean area, where the biogeochemical processes of Se are not well known. The four soils used were characterized by different physicochemical characteristics (Table 9.2). The G soil, Typic Hapludalf, was collected in Greece, the NI soil, Typic Eutrochrept, in Northern Italy, the CI, Eutrochreptic Rendoll, in Central Italy, and the S soil, Entic Hapludoll, in Spain. The taxonomic classification of the soils followed the United States Department of Agriculture nomenclature (USDA, 1985). The isotherms of selenate sorption were obtained by shaking 2.5 g of soil with 25 mL of solution prepared from Na_2SeO_4 with distilled water (pH 5.3) and containing 1, 3, 5, 10, 20, 40, 60, 80, or 100 µg Se g^{-1}. Suspensions were centrifuged, the supernatants were filtered, and the samples were then acidified with concentrated HCl in order to analyze the Se. Desorption isotherms were determined by resuspending the samples in phosphate solution (0.1 M KH_2PO_4 in distilled water, pH 5.3) under the same experimental conditions

Table 9.2   Main Physicochemical Characteristics
            of the Soils Used

		G	CI	NI	S
Clay	(%)	7.65	21.0	7.10	12.5
Silt	(%)	14.0	36.4	24.8	28.2
Sand	(%)	78.3	42.6	67.6	59.3
Organic matter	(%)	0.52	1.59	5.83	10.7
pH (H$_2$O)		8.0	7.5	4.5	7.4
CEC	(meq/100g)	11.25	14.37	30.6	27.5
CaCO$_3$	%	2.19	3.21	—	12.9

Table 9.3   Freundlich Constants
            for Selenate Sorption

	G	CI	NI	S
K	3.74	6.10	9.78	15.96
n	1.35	1.44	1.51	1.84
r	0.978	0.981	0.996	0.979

r = regression coefficient

as for the selenate sorption. The suspensions were then filtered, and the filtrates were analyzed for Se. Atomic absorption spectroscopy with hydride generation was used to analyze selenium. The sorption data were analyzed by the Freundlich equation:

$$q = KC^{1/n}$$

where q is the amount of selenium sorbed on unit weight of soil ($\mu g\ g^{-1}$), K and $1/n$ are empirical parameters related to the sorption capacity, and C is the equilibrium selenium concentration ($\mu g\ ml^{-1}$). The parameters of the Freundlich equation are reported in Table 9.3. The four soils varied considerably in their ability to sorb added selenate, and the isotherm patterns were significantly different in the four soils. The Se sorption by the soils from Northern Italy (NI) and Spain (S), characterized by high CEC values and high organic matter content, can be described by an L-type curve (Figure 9.1). They showed a great affinity for selenium, as indicated by the higher K values. In the soil from Spain, the high CaCO$_3$ content might have increased the selenate sorption in spite of the high pH values (Singh et al., 1981). The reactive levels of Ca, in fact, control the sorption process of selenium in calcareous soils, the processes responsible for Se sorption on carbonate surfaces in soils being ligand exchange or chemisorption. In the soils from Greece (G) and Central Italy (CI), the isotherms showed quite a different isotherm pattern, and the sorption can be described by an S-type curve (Figure 9.2). The K parameter of the Freundlich equation showed lower values, indicating a reduced sorption capacity of these soils and reflecting the number of sites on the soil surfaces involved in the sorption process. The organic matter content played an important part in the adsorption of selenium, in agreement with earlier investigations which showed that the highest

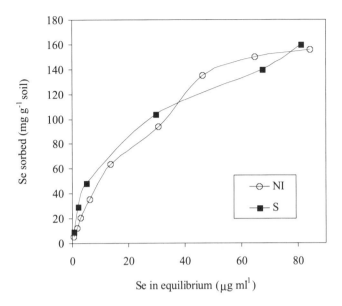

Figure 9.1 Equilibrium sorption isotherms for selenate on soils from Northern Italy (NI) and Spain (S).

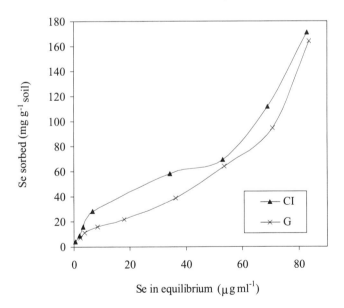

Figure 9.2 Equilibrium sorption isotherms for selenate on soils from Central Italy (CI) and Greece (G).

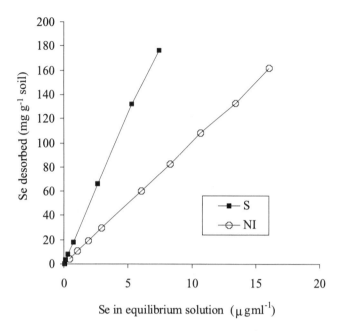

Figure 9.3   Equilibrium desorption isotherms for selenate on soils from Northern Italy (NI) and
Spain (S).

amount of selenium was retained by soils with high organic matter content (Singh
and Singh, 1979). A limited interaction between organic matter and selenate ions
could be expected due to the similarity of the electrical charges. Selenate, similar
to phosphate (Parfin, 1978), could be sorbed by iron and aluminum ions, which can
also be chelated by humic organic molecules in soils.

Selenate sorption by soil seems not to be affected by pH, as reported in previous
investigations (John et al., 1976). The lowest Se sorption by the Greek soil, could
be ascribed either to the low organic matter content and to the high pH value, which
maintained the solubility of selenium.

Desorption experiments showed that the amount of desorbed selenate varied
among the investigated soils and that the isotherms were linear, indicating that the
greater the amount sorbed, the greater the desorption (Figures 9.3 and 9.4). The
differences between the desorption and the sorption isotherms might suggest the
existence of hysteresis or irreversible sorption, as a result of a permanent change in
the soil-selenate system following the sorption process.

Phosphate desorbed Se into the solution phase, but soil irreversibly retained a
significant proportion of selenium, suggesting that there might be linkages which
allow the release of selenium into the soil solution only after physicochemical
variation, such as the exchange with phosphate ions. The fraction that is resistant
to extraction by phosphate solution represents a substantial reservoir of Se in the
soil which may be resistant to leaching.

The highest value of selenate sorption was determined in the Spanish soil from
which the selenate was released quite easily. Selenate can thus be sorbed by ferric
and aluminum ions complexed by organic matter, but they are easily available for

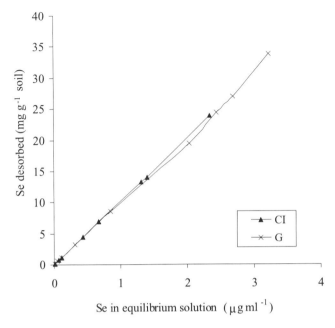

Figure 9.4   Equilibrium desorption isotherms for selenate on soils from Central Italy (CI) and
Greece (G).

exchange with the added phosphate ions. The Greek soil, which showed the lowest
sorption capacity, firmly retained the sorbed selenium, probably because only high-
energy sites were involved in the sorption process.

From the data obtained in this study, it appears that at lower levels of selenate
in the soil-solution system, more sites were available for high affinity sorption, and
selenium was firmly retained. When increasing concentrations of selenate are added,
fewer affinity sites were involved in the sorption process and selenium was more
easily removed by the phosphate solution.

With high concentrations of Se, the soil retention is reduced and the bioavailability of selenium in the environment might increase and induce groundwater and
food chain contamination.

Our studies emphasize the important role of soil organic matter in the selenate
sorption process. This is significant, as interest in the study of selenium develops
particularly when biomasses from different sources are added to soil. This practice
not only introduces increasing quantities of selenium into the soil, but also relevant
amounts of organic materials, which can naturally influence the dynamics of the
elements in the soil.

The processes of sorption and desorption are of primary importance in order to
understand the behavior of selenium in soil. These studies should be considered as
a starting point for the further detailed study of the biogeochemical processes of Se
in the soils of the Mediterranean area. They are also important in determining the
links between sorption-desorption processes and bioavailability in order to quantify
the toxicological hazard of Se in soil.

REFERENCES

Abrahms, M.A., C. Shennan, R.J. Zasoski, R.G. Burau. 1990. Selenomethionine uptake by wheat seedlings. *Agron. J.* 82, pp.1127.

Atkins, C.E., E. Epstein, R.G. Burau. 1988. Absorption and transport of selenium by plant. In *Selenium Contents in Animal and Human Food Crops Grown in California*, Tanji, K.K., Ed., pp. 29-32.

Balistrieri, L.S. and T.T. Chao. 1987. Selenium adsorption by goethite. *Soil Sci. Am. J.* 51, pp. 1145-1151.

Balistrieri, L.S. and T.T. Chao. 1990. Adsorption of selenium by amorphous iron oxyhydroxide and manganese dioxide. *Geochim. Cosmochim Acta.* 54, pp. 739-751.

Bar-Yosef, B. 1987. Selenium desorption from Ca-kaolinite. *Comm. Soil Sci. Plant Anal.* 18, 7, pp. 771-779.

Banuelos, G.S. and D.W. Meek. 1990. Accumulation of selenium in plants grown on selenium-treated soil. *J. Environ. Qual.* 19, pp. 772-770.

Bell, P.F., D.R. Parker, A. Page. 1992. Contrasting selenate-sulfate interactions in selenium-accumulating and non-accumulating plant species. *Soil Sci. Am. J.* 56, pp. 1818-1824.

Carter, D.L., C.W. Robbins, M.J. Brown. 1972. Effects of phosphorus fertilization on the selenium concentration in alfalfa. *Soil Sci. Soc. Am. Proc.* 36, pp. 624-628.

Chortyk, O.T, J.F. Chaplin, W.S. Schlotzhaver. 1984. Growing selenium-enriched tobacco. *J. Agric. Food Chem.* 32, pp. 64-68.

Cowan, E.C., J.M. Zachara, C.T Resch. 1990. Solution ion effects on the surface exchange of selenite on calcite. *Geochim. Cosmochim. Acta.* 54, pp. 2223-2234.

El Rashidi, M.A., D.C. Adriano, S.M. Workman, W.L. Lindsay. 1987. Chemical equilibria of selenium in soils-a theoretical development. *Soil Sci.* 144, pp. 141-151.

Fio, J.L., R. Fujii, S.J. Deverel. 1991. Selenium mobility and distribution in irrigated and nonirrigated alluvial soils. *Soil Sci. Am. J.* 55, pp. 1313-1320.

Glasauer, S., H.E. Doner, A.U. Gehring. 1995. Sorption of selenite to goethite in a flow-through reaction chamber. *Eur. J. Soil Sci.* 46, pp. 47-52.

Goldberg, S. and R.A. Glaubig. 1988. Anion sorption on a calcareous, montmorillonitic soil-selenium. *Soil Sci. Am. J.* 52, pp. 954-958.

Gupta, U.C., K.B. McRae, K. A. Winter. 1982. Effect of applied selenium on the selenium content of barley and forages and soil selenium depletion rates. *Can. J. Soil. Sci.* 62, pp. 145-154.

Gustafsson, J.P. and L. Johnsson. 1992. Selenium retention in the organic matter of Swedish forest soils. *J. Soil. Sci.* 43, pp. 461-472.

Hayes, K.F., P. Charalambos, J.O. Leckie. 1992. Modelling ionic strength effects on anion sorption at hydrous oxide/solution interfaces. *J. Colloid Interface Sci.* 25, pp. 717-726.

Hingston, F.J., A.M. Posner, J.P. Quirk. 1972. Anion sorption by goethite and gibbsite I. The role of the proton in determining sorption envelopes. *Soil Sci. Am. J.* 23, pp. 177-192.

Hopper, J.L. and D.R. Parker. 1999. Plant availability of selenite and selenate as influenced by the competing ions phosphate and sulfate. *Plant and Soil,* 210, pp.199-207.

Jayaweera, G.R., and J.W. Biggar. 1996. Role of redox potential in chemical transformations of selenium in soils. *Soil Sci. Am. J.* 60, pp. 1056-1063.

John, M.K., W.M. Saunders, J.H. Watkinson. 1976. Selenium adsorption by new Zealand soils *N. Z. J. Agr. Res.* 19, pp. 143-151.

Johnsson, L. 1989. Se levels in the mor layer of Swedish forest soils. *J. Soil Sci.* 43, 3, pp. 461-472.

Lakin, H.W. 1973. Selenium in our environment, in *Trace Elements in the Environment.* Adv. Chem. Ser. 123, pp. 96-111.

Lewis, B.G. 1976. Selenium in biological systems, and pathways for its volatilization in higher plants, in *Environmental Biogeochemistry*, Nriagu, J.O., Ed., Ann Arbor Science, Ann Arbor, MI, pp. 389-409.

Mayland H.F., L.F. James, K.E. Panter, J.L. Sonderegger. 1989. Selenium in seleniferous environments. In *Selenium in Agriculture and the Environment*, Jacobs, L.W., Ed., SSSA Special Publication 23, American Society of Agronomy, Inc., pp. 15-50.

Mayland, H.F. 1994. Selenium in plant and animal nutrition. In *Selenium in the Environment*, Frankenberger W.T. Jr. and Benson S., Eds., Marcel Dekker, Inc., New York.

McNeal, J.M. and L.S. Balistrieri, 1989. Geochemistry and occurrence of selenium: an overview. In *Selenium in Agriculture and the Environment*, Jacobs, L.W., Ed., SSSA Special Publication.

Mikkelsen, R.L, F.T. Bingham, A.L. Page. 1989. Factors affecting selenium accumulation by agricultural crops. In *Selenium in Agriculture and the Environment*, Jacobs, L.W., Ed., SSSA Special Publication 23, American Society of Agronomy, Inc., pp. 65-95.

Neal, R.H. 1995. Selenium. In *Heavy Metals in Soils*, Alloway, B., Ed., Blackwell Publishers, Oxford, U.K.

Neal, R.H., G. Sposito, K.M. Holtzclaw, S.J. Traina. 1987. Selenite adsorption on alluvial soils: I. Soil composition and pH effects. *Soil Sci. Am. J.* 51, pp. 1161-1165.

Neal, R.H. and G. Sposito. 1989. Selenate adsorption on alluvial soils. *Soil Sci. Am. J.* 53, pp. 70-74.

Parfin R.L. 1978. Anion adsorption by soils and soil material. *Adv. Agron.* 30, pp. 1-50.

Pezzarossa, B., G. Petruzzelli, D. Piccotino, F. Malorgio, C. Hillhouse, J. Jones, C. Shennan. 1997. Selenium uptake and partitioning in tomato plants in relation to sulfate concentration in soil, in *Contaminated Soils: Third International Conference on the Biogeochemistry of Trace Elements, Paris*, May 15-19, 1995, Prost, R., Ed., INRA Editions, Paris, France, D:\data\communic\056.PDF, Colloque 85.

Pezzarossa, B., D. Piccotino, C. Shennan, F. Malorgio. 1999. Uptake and distribution of selenium in tomato plants as affected by genotype and sulfate supply. *J. Plant Nutr.,* 22 (10), pp. 1613-1635.

Rajan, S.S.S. and J.W. Watkinson. 1979. Sorption of selenite and phosphate onto allophane clay. *Soil Sci. Am. J.* 40, pp. 51-54.

Rosenfeld, I. and Beath O.A. 1964. *Selenium: Geobotany, Biochemistry, Toxicity and Nutrition.* Academic Press, New York.

Saeki, K., S. Matsumoto, R. Tatsukawa. 1995. Selenite sorption by manganese oxides. *Soil Sci.* 160, pp. 265-272.

Sarathchandra, S.U. and J.H. Watkinson. 1981. Oxidation of elemental selenium to selenite by *Bacillus megaterium. Science,* 21, 600-601.

Shrift A. 1973. Metabolism of selenium by plants and microorganisms. In *Organic Selenium Compounds: Their Chemistry and Biology*, Klayman, D.L. and W.H. Gunther, Eds., John Wiley & Sons, New York, pp. 763-814.

Singh, M. and N. Singh. 1979. The effect of forms of selenium on the accumulation of Se, S, and forms of N and P in forage cowpea. *Soil Sci.* 12, pp. 264-269.

Singh, M., N. Singh, P.S. Relan. 1981. Adsorption and desorption of selenite and selenate selenium on different soils. *Soil Sci.* 132, pp. 134-141.

Sylvester, M.A. 1986. Results of U.S. Geological Survey studies pertaining to the agricultural drainage problem of the western San Joaquin Valley, California. In *Selenium and Agricultural Drainage: Implications for San Francisco Bay and the California Environment*, Proc. 2nd Selenium Symp., Berkeley, CA, 23 Mar. 1985, The Bay Institute of San Francisco, Tiburon, CA, pp. 35-40, 1986.

Tan, J.A., W.Y. Wang, D.C. Wang, S.F. Hou. 1994. Adsorption, volatilization, and speciation of selenium in different types of soils in China. In *Selenium in the Environment*, Frankenberger W.T. Jr. and Benson S. Eds., Marcel Dekker, Inc., New York, pp. 47-67.

Underwood, E.J. 1977. Selenium, in *Trace Elements in Human and Animal Nutrition*, 4th ed., Academic Press, New York, pp. 302-346.

United States Department of Agriculture. 1985. *Keys to Soil Taxonomy*. Management Support Service. Technical Monograph 6, USDA Publ. Cornell Univ., Ithaca, New York.

Vokal-Borek, H. 1979. *Selenium*. USIP REPORT 79-16, University of Stockholm, Institute of Physics.

Yang, G., S. Wang, R. Zhou, S. Sun. 1983. Endemic selenium intoxication of humans in China. *Am. J. Clin. Nutr.,* 37, pp. 872-881.

CHAPTER 10

Arsenic Behavior in Contaminated Soils: Mobility and Speciation

Virginie Matera and Isabelle Le Hécho

INTRODUCTION

Due to manufacture of arsenic-based compounds, smelting of arsenic-containing ores, and combustion of fossil fuels, arsenic is introduced into soils, waters, and the atmosphere (Azcue et al., 1994). The natural content of arsenic in soils is 5 mg kg^{-1} (Backer and Chesnin, 1975). The occurrence of arsenic in the environment may be due to both background and anthropogenic sources. In the first case, arsenic is concentrated in magmatic sulfides and iron ores. The most important arsenic ores are arsenic pyrite or mispickel (FeAsS), realgar (AsS), and orpiment (As$_2$S$_3$). Human activities may lead to arsenic accumulation in soils mainly through use or production of arsenical pesticides (fungicides, herbicides, and insecticides). Arsenic is a contaminant that represents a potential risk for man, especially in mining districts and near active smelters, by ingestion/inhalation of arsenic-bearing particles. Arsenic is also phytotoxic: an average toxicity threshold of 40 mg kg^{-1} has been established for crop plants (Sheppard, 1992).

To prevent As toxicity and to access the contamination risk of the environment, numerous reviews have been published in recent years describing the behavior, chemistry, and sources of arsenic in the soil environment (Sadiq, 1997; Smith et al., 1998). Furthermore, many previous studies have investigated arsenic sorption on well-characterized solid phases (Pierce and Moore, 1982; Sun and Doner, 1996; Manning and Goldberg, 1997a; Frost and Griffin, 1977; Goldberg and Glaubig, 1988). Work done on historically contaminated soils consist mainly of spatial distribution (Lund and Fobian, 1991; Sadler et al., 1994; Voigt et al., 1996), determination of arsenic or of the different parameters able to influence arsenic mobilization

(Masscheleyn et al., 1991; Pantsar-Kallio and Manninen, 1997; Davis et al., 1994), or arsenic-bearing phase determination (Davis et al., 1996; Juillot et al., 1999).

Some of the research works that investigate arsenic mobility in historically contaminated soils follow a global reasoning, showing detailed characterizations of the soils as a means to understanding the nature of the arsenic-bearing phases. A better knowledge of arsenic-bearing phases in relation to speciation and mobilization will help to better manage arsenic-polluted soils. The purpose of this chapter is to report the characterization of arsenic-bearing phases resulting from a historically polluted soil. Geological studies of the site investigated in this work have been done (Piantone et al., 1994; Braux et al., 1993). The soil is collected from a former gold mine heavily polluted by arsenic due to anthropogenic sources: pyrite and arsenopyrite oxidation. On this site, mining activities started in the beginning of the century and ceased in the 1950s.

The arsenic-bearing phase determination was done using three different but complementary speciation methods:

- Analytical chemical speciation (HPLC-ICP-MS)
- Localization phase speciation (sequential extractions)
- Physical speciation (SEM, XRD)

This characterization is an essential step toward a better understanding of arsenic forms and arsenic mobilization mechanisms of this historically contaminated soil.

Batch experiments and column transport experiments using small saturated columns were done to investigate arsenic remobilization under the influence of different physicochemical parameters (pH and phosphate concentrations). Before the presentation of the different experimental results, a literature study summarizes arsenic geochemistry in contaminated soils.

ARSENIC GEOCHEMISTRY IN CONTAMINATED SOILS

Arsenic Chemistry in Soils

Arsenic (atomic number 33; atomic mass 74.9216) has an outer electron configuration of $4s^2 4p^3$ and belongs to subgroup V of the Periodic Table. It is often described as a metalloid. In soils, the chemical behavior of arsenic (As) is, in many ways, similar to that of phosphorus (P).

Because the solubility, mobility, bioavailability, and toxicity of As depend on its oxidation state (Masscheleyn et al., 1991), studies of As speciation and transformation among species are essential to understanding As behavior in the environment. The rate of As transfer is not only a function of As concentration in soil, but is also largely influenced by its geochemical behavior. Important factors affecting As chemistry in soils are soil solution chemistry, solid phase formation, adsorption and desorption, effect of redox conditions, biological transformations, volatilization, and cycling of As in soils (Sadiq, 1997).

Arsenic Speciation in Soils and Porewaters

In natural systems, As may occur in four oxidation states: (−3), (0), (+3), and (+5). Arsenate (As(V)) and arsenite (As(III)) are the main forms in soils (Harper and Haswell, 1988) even if we sometimes may expect to find the oxidation states (−3) and (0) in very highly reducing conditions (McBride, 1994). Nearly 90% of the species of As in aerobic soils — in mineralized areas or not — are arsenates, whereas only 15 to 40% of As is found under the oxidation state (+5) in soils saturated with water in anaerobic conditions (O'Neill, 1995). The potential mobility (i.e., solubility) of As is based on these oxidation states. For example, As(V) is less toxic than As(III) (Ferguson and Gavis, 1972); As(V) sorbs more strongly than As(III) (Pierce and Moore, 1982); As(III) is more soluble and mobile than As(V) (Deuel and Swoboda, 1972). In general, As(V) compounds predominate in aerobic soils, whereas As(III) compounds predominate in slightly reduced soils. As also appears to be more mobile under both alkaline and more saline conditions.

The changes in the oxidation states linked to the variations in pH and Eh have slow kinetics in an aqueous system, which explains why the species found in interstitial waters do not always follow to the expected distribution. McGeehan and Naylor (1994) show that rates of desorption and disappearance of H_3AsO_3 and $H_2AsO_4^-$ are slower in soil with higher adsorption capacity, suggesting that sorption processes may influence redox transformations of As oxyanions.

Inorganic Arsenic

Arsenic ionized species are mainly oxyanions which exhibit various degrees of protonation and valence charge, depending on pH. O'Neill (1995) gives the balanced solution of arsenous acid (As III) and arsenic acid (AsV). Arsenite (As III) can appear in the forms: H_3AsO_3, $H_2AsO_3^-$, $HAsO_3^{2-}$, and AsO_3^{3-}; arsenate (As V), mainly in the forms: H_3AsO_4, $H_2AsO_4^-$, $HAsO_4^{2-}$ and AsO_4^{3-}. The pKa values indicate that predominant arsenic species for $2 < pH < 9$ are:

- H_3AsO_4 for As (III)
- $H_2AsO_4^-$ and $HAsO_4^{2-}$ for As (V)

Arsenic in primary minerals is found in four oxidation states: (−III) (arsenides and gaseous compounds such as arsine [AsH_3] and arsenic chloride [$AsCl_3$]), (0) (native arsenic), (+III) (oxides, sulfides, sulfosalts, and arsenites), and (+V) (arsenates) (Escobar Gonzales and Monhemius, 1988).

- Arsenide minerals are important in the extractive metallurgy of cobalt, nickel, platinum, palladium, iridium, and ruthenium. Among these, the cobalt arsenides (skutterudite [$CoAs_3$]) and nickel arsenides (niccolite [$NiAs$] and rammelsbergite [$NiAs_2$]) are the most abundant. With antimony, As forms minerals such as allemontite ($AsSb$). Also found are loellingite ($FeAs_2$) and domeykite (Cu_3As).
- Sulfur minerals (sulfides and sulfosalts) are stable under reducing conditions. Main species that commonly occur in the environment are: arsenopyrite ($FeAsS$), orpiment (As_2S_3), realgar (As_4S_4), enargite (Cu_3AsS_4), colbaltite ($CoAsS$), and proustite (Ag_3AsS_3). Realgar is currently found as a minor constituent of certain ore veins.

Orpiment and realgar may have been formed during oxidation processes, mainly of arsenopyrite.

- Arsenite minerals (armangite ($Mn_3(AsO_3)_2$, finnemanite ($Pb_5(AsO_3)_3Cl$), and reinerite ($Zn_3(AsO_3)_2$)) are found in endogenous deposits and have only a restricted range of thermodynamic stability. Few of these minerals are present in soil.
- Oxides are formed at high temperature (for example: claudetite/arsenolite (As_2O_3)) and are rare due to their high solubility in water.
- Origin of arsenates has not been defined. They may have been formed *in situ* as products of the oxidation of arsenides and sulfoarsenides. Alternatively, formation may have been due to decomposition of arsenides or sulfoarsenides, followed by dissolution and transport of As with eventual reprecipitation elsewhere as an arsenate mineral. In these two cases, these processes lead, by precipitation, to numerous arsenate formations. In nature, arsenates of Al, Bi, Be, Ca, Cu, Co, Fe, Hg, Mn, Mg, Ni, Pb, Zn, and U have been encountered; however, only the arsenates of Ca, Fe, Mn and Pb are abundant. Arsenates usually found in the environment are: scorodite ($FeAsO_4 \cdot 2H_2O$), pharmacosiderite ($Fe_4(AsO_4)_3(OH)_3 \cdot 6H_2O$), parasymplesite/symplesite ($Fe_3(AsO_4)_2 \cdot 8H_2O$), pharmacolite ($CaHAsO_4$), erythrite ($Co_3(AsO_4)_2 \cdot 8H_2O$), and annabergite ($Ni_3(AsO_4)_2 \cdot 8H_2O$).

Organic Arsenic

The organic forms of As are often linked to methylation reactions by microorganisms. Methylation of oxyanions leads to the formation of compounds such as (O'Neill, 1995):

- Monomethylarsonic acid (MMAA) $CH_3AsO(OH)_2$ (and monomethylarsenate [MMA]: MMAA salt)
- Dimethylarsinic acid (DMAA) (cacodylic acid) $(CH_3)_2AsO(OH)$ (and dimethylarsenate [DMA]: DMAA salt)
- Trimethylarsenic oxide $(CH_3)_3AsO$
- Dimethylarsine $(CH_3)_2AsH$
- Trimethylarsine $(CH_3)_3As$

The biomethylation reactions depend upon the microorganisms and the As compounds over a wide range of pH conditions, whereas many other microorganisms appear much more limited in the substrates they can methylate and the degree of methylation they can produce (O'Neill, 1995). The presence of such compounds in soil can be linked to the supply of anthropogenic compounds, such as fertilizers and pesticides.

Phenomena Affecting Arsenic Mobility in Soils

As mobilization in soils depends on different processes: oxidation/reduction; complexation/coprecipitation; adsorption/desorption, and As-bearing phases (soil properties). One of the most commonly reported, and perhaps the first reaction to occur in soils, is As adsorption on soil particles. Numerous studies have dealt with As sorption on specific minerals and on uncontaminated soils. The soil properties reported to be most related to As sorption are: iron, aluminum, and manganese (hydr)oxides (Pierce and Moore, 1980; Pierce and Moore, 1982; Oscarson et al.,

1981), clay content (Manning and Goldberg, 1997a; Frost and Griffin, 1977; Xu et al., 1991), and organic matter (Lund and Fobian, 1991; Thanabalasingam and Pickering, 1986). pH and Eh are factors usually studied in these works. Coprecipitation and adsorption of As with iron oxides may be the most common mechanism affecting its mobility under most environmental conditions. In addition to adsorption, As(V) and As(III) species also can be removed from minerals by substitution with phosphate.

Oxidation and Reduction

Masscheleyn et al. (1991) showed that solubility and speciation of As in soils is governed mainly by redox potential. Under oxidizing conditions (200 to 500 mV), As solubility is low and most (65 to 98%) is present as As(V). Under moderately reducing conditions (0 to 100 mV), As solubility is controlled by iron oxyhydroxides. Arsenate is coprecipitated with iron compounds and released upon solubilization. If strong reducing conditions dominate (-200 mV), which corresponds to flooded soils, soluble As increases 13-fold over 500 mV redox. Sadiq (1997) mentions the importance of sulfur on As mobility. In soils with redox conditions more oxidized than pe+pH $\geq$ 5, $Fe_3(AsO_4)_2$ is more stable than all As(III) minerals, whereas, in more reduced soils, sulfides of As(III) are the most stable As minerals. In anoxic soil systems, arsenous oxides are less stable than sulfides.

Transformation kinetics of As (V) to As(III) are very slow, which explains that an important amount of As (V) can be found under strong reducing conditions (Onken and Hossner, 1996). Transformation between the various oxidation states and species of As may occur as a result of biotic or abiotic processes (McGeehan and Naylor, 1994; Masscheleyn et al., 1991).

Thus, in some aquatic sediments and in soils, H_3AsO_3 is easily oxidized in $H_2AsO_4^-$ through abiotic processes. Oscarson et al. (1981), showed that this oxidation is catalyzed by Mn dioxides present in sediments, whereas Fe(III) oxide occurrence cannot manage As (III) transformation to As (V). Moreover, bacterial oxidation of H_3AsO_3 to $H_2AsO_4^-$ has been observed in mine waters (Wakao et al., 1988). Biotic reduction of $H_2AsO_4^-$ has been observed in groundwater (Agget and O'Brien, 1985) and soils (Cheng and Focht, 1979).

Complexation and Precipitation

Because of similarity in the nature of charges on both organic molecules and As chemical forms, As has demonstrated a limited affinity for organic complexation in soil. However, systematic field information on the occurrence and persistence of organic As complexes in soil solutions is limited. It is generally accepted that organoarsenical complexes constitute a minor fraction of total dissolved As in soil solutions (Sadiq, 1997). Sadiq et al. (1983) found that in well-oxidized and alkaline soils, occurrence of As and major elements (Ca, Mn, Mg, etc.) can form precipitates such as $Ca_3(AsO_4)_2$, which is the most stable As mineral, followed by $Mn_3(AsO_4)_2$.

In soils with high sulfate concentrations and under reducing conditions, As and sulfur ($-II$) can form very insoluble compounds such as arsenopyrite (Gustafsson and Tin, 1994). Solubility of these precipitates depends on the oxidation state of As and on pH conditions. Solubility of As(V)/Fe(III) precipitates decreases when pH decreases, whereas solubility of As(III)/Fe(III) decreases when pH increases (Gulens et al., 1979). However, according to Livesey and Huang (1981), arsenic retention by soils does not proceed through the precipitation of sparingly soluble arsenate compounds. Arsenate retention evidently proceeds through the adsorption mechanism.

Adsorption and Desorption

Arsenic mobilization is mainly controlled by adsorption/desorption processes. Studies of As sorption are carried out with different procedures and on various matrices. In general, these phenomena are linked mainly to pH and also to redox conditions, mineral nature, and As state oxidation. The surface charge properties of soil are strongly influenced by soil pH. Acid soils have large amounts of positive charges, and adsorption of the $H_2AsO_4^-$ anion may become important. Arsenate anions are attracted to positively charged colloid surfaces either at broken clay lattice edges where charged Al^{3+} groups are exposed, or on the surfaces of iron and aluminum oxides and hydroxide films (Brookins, 1988). As(III) and As(V) adsorption has been studied according to soil constituent nature. Some of these studies are presented below:

Clay Minerals

The amount of anion adsorption by clay minerals is usually small compared to the amount of cation exchange adsorption. Anion adsorption sites on clay particles are associated with exposed octahedral cations on broken clay particle edges (Van Olphen, 1963). These mineral phases can contribute to As adsorption through surface reactions such as Reaction 10.1, where M is an exposed octahedral cation (Frost and Griffin, 1977; Manful et al., 1989):

$$-M-OH + H_2AsO_4^- \Leftrightarrow -M-H_2AsO_4 + OH^- \qquad (10.1)$$

Thus, on a simple mass action basis, the extent of surface activation will depend upon solution pH. More recently, Lin and Puls (2000), studied adsorption, desorption, and oxidation of As affected by clay minerals. They found that, in general, the clay minerals exhibited less As(III) adsorption than As(V) adsorption, and they confirmed that adsorption was affected by pH and that arsenate sorption on clay minerals can occur through edge defects (e.g., protonation of broken Al-OH bonds exposed at particle edges), but they also suggested that at high loadings of arsenate, arsenate sorption on halloysite (1:1 layer clay) may be controlled by the formation of a hydroxy-arsenate interlayer, which may be more important to As(V) adsorption than adsorption with the surface hydroxyl groups.

Arsenate sorption on the clay minerals kaolinite and montmorillonite increases below pH 4, exhibits a peak between pH 4 and 6, and decreases above pH 6; arsenite sorption on montmorillonite peaks near pH 7, while arsenite sorption on kaolinite increases steadily from pH 4 to 9 (Frost and Griffin, 1977). Moreover, Xu et al.

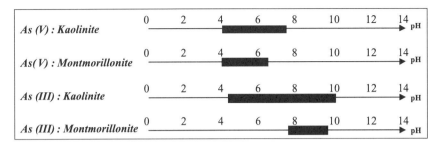

Figure 10.1 Adsorption maximum of arsenates and arsenites on clay minerals. (Data from Manning and Goldberg, 1997a; Goldberg and Glaubig, 1988; Frost and Griffin, 1977.)

Figure 10.2 Adsorption maximum of arsenates on calcite. (Data from Goldberg S. and Glaubig R.A., 1988. Anion sorption on a calcareous, montmorillonitic soil-arsenic, *Soil Sci. Soc. Am. J.*, 2, pp. 1297-1300.)

(1988) studied As(V) adsorption on kaolin and alumina. As(V) adsorption had a maximum around pH 5 and decreased drastically above pH above 6.

Arsenite adsorption on montmorillonite (Figure 10.1) increases with pH to reach an adsorption maximum near pH 7. On kaolinite (Figure 10.1), adsorption increases steadily from pH 4 to 9 (Frost and Griffin, 1977; Goldberg and Glaubig, 1988). Manning and Goldberg (1997a) found the same trends, however, As(III) adsorption on kaolinite shows an adsorption maximum around pH 9, whereas, on montmorillonite, maximum is reach around pH 8.

Calcite or Calcium Saturated Soils

The adsorption of As (V) on calcite increases from pH 6 to 10. It presents a maximum between pH 10 and 12 and then it decreases (Figure 10.2). Such studies could not be carried out on As(III) (Goldberg and Glaubig, 1988). Brannon and Patrick (1987) found a correlation between sorbed As and the calcium carbonate content in the case of sediments. They suggest the possibility of the carbonates being covered by iron oxides or aluminum hydrated oxides. So, the role of the carbonates or calcite would not seem as evident as the part played by iron oxides or aluminum oxides in As sorption. Otherwise, since calcium arsenate is more soluble than aluminum and iron arsenates, the calcium influence would prove to be less important than the iron or aluminum.

Oxides and Hydroxides

All the studies underscore the critical role of environmental pH. Pierce and Moore (1980) show that As(V) has an adsorption maximum at pH 4, where $H_2AsO_4^-$ is the dominant species, and that As(III) in the form of H_3AsO_3 reaches this maximum

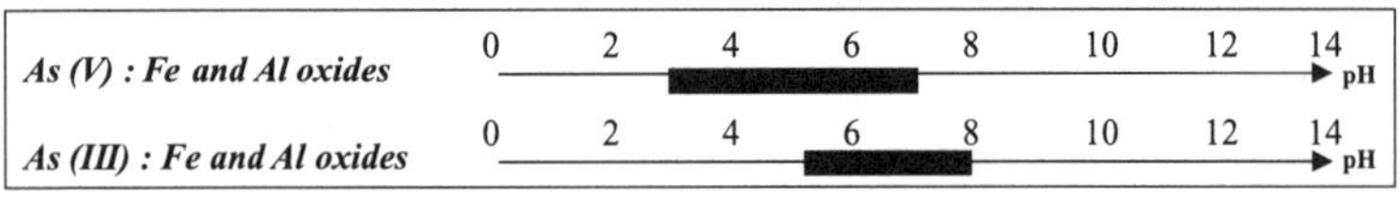

	0	2	4	6	8	10	12	14
As (V) : Fe and Al oxides								pH
	0	2	4	6	8	10	12	14
As (III) : Fe and Al oxides								pH

Figure 10.3   Adsorption maximum of arsenates and arsenites on iron and aluminum oxides (Data from Pierce and Moore, 1982; Xu et al., 1988; Gupta and Chen, 1978; Anderson et al., 1975; Dzombak and Morel, 1990.)

at pH 7. Moreover, arsenate sorption would seem to be more important than arsenite sorption in the case of iron or aluminum hydroxide matrix. Several studies (Figure 10.3) confirm that As(III) has a sorption maximum at pH 7 to 8 (Xu et al., 1988; Pierce and Moore, 1982). For pentavalent As, sorption reaches a maximum around pH 4 or 5 and then decreases with more alkaline pH (Xu et al., 1988; Pierce and Moore, 1982; Gupta and Chen, 1978; Anderson et al., 1975). Manning and Goldberg (1997b) also show that As(V) reaches sorption maximum at pH 4 to 7, then decreases at more alkaline pH levels. The capacity of manganese oxides, for adsorbing As(III) or As(V) would seem to depend, in part, on the point of zero charge (PZC) of these solids. Since birnessite (δ-MnO$_2$) pH$_{zpc}$ is low, As sorption, mainly in oxyanion forms, is not favored because of the relatively high energy barrier (Oscarson et al., 1981).

Organic Matter

Thanabalasingam and Pickering (1986) studied the adsorption of As(III) and As(V) on humic acids. They showed that adsorption depended on pH but also on humic acid content. For As(V), the maximum adsorption is around pH 5.5 and it occurs at higher pH for As(III). For more acid pH, sorption decreases. The more alkaline the pH, the more soluble the humic substances, and their capacity to retain As is reduced. Humic acids can be an important factor in As adsorption in relatively acid environments; on the other hand, alkaline conditions contribute to release of As.

Xu et al. (1988) showed that a low concentration of fulvic acids leads to an appreciable reduction of As adsorption on alumina. These organic acids compete with As for adsorption sites. Moreover, this effect is minor under acid conditions (pH 3) as well as alkaline conditions (pH 9) (Xu et al., 1991).

Phosphates and Other Competitive Anions

Phosphorus and As both form oxyanions in the (+5) oxidation state. Phosphates are stable over a large range of pH and Eh, while As can exist in the (+3) oxidation state and easily forms links with S and C (O'Neill, 1995). Thus, phosphates strongly compete with As for adsorption sites in environmentally important pH ranges. Phosphates are able to limit As adsorption by humic substances since 60% of adsorbed As(V) and 70% of adsorbed As(III) were desorbed by H$_2$PO$_4$ into a 10^{-6} M phosphate solution (Thanabalasingam and Pickering, 1986). Bhumbla and Keefer (1994) emphasized the strong adsorption of phosphate on amorphous oxides. They also showed that phosphates have a better affinity for aluminum oxides than As.

Similarly, phosphate ions substantially suppressed the sorption of $H_2AsO_4^-$ but not proportionally with increasing phosphate/arsenate molar ratio (Manful et al., 1989).

Competition for the sorption sites by Cl^-, NO_3^- and SO_4^{2-} ions were less than PO_4^{3-}. Hence, these ions could not significantly suppress the sorption of arsenate ions (Manful et al., 1989). Nevertheless, Xu et al. (1988) showed that when pH is less than 7, the presence of sulfate causes a decrease in the adsorption of As(V) on alumina. However, when the sulfate concentration is further increased (from 20 to 80 mg l^{-1}), the difference in adsorption is not significant. This observation suggests that sulfate can compete with $H_2AsO_4^-$ and $HAsO_4^{2-}$ and occupy surface sites on the alumina. Furthermore, the ionic strength would seem to have an influence on As mobilization. Pantsar-Kallio and Manninen (1997) have tested different solutions for As desorption. The efficiency of the different solutions follows the order: Na_2CO_3 > $NaHCO_3$ > K_2SO_4 > $NaNO_3$. Amounts of As extracted also increased with increasing carbonate concentrations.

Arsenic-Bearing Phases in Mine Sites

Some studies reported that, in As-contaminated soils, the primary As mineral assemblage may weather, resulting in formation of a secondary mineral. Secondary precipitation of As compounds may occur on soil colloid surfaces subsequent to its adsorption, and direct precipitation of As solid phases may occur (Sadiq, 1997). When released in the environment, As can be incorporable into various As-bearing minerals. For example, in contaminated soils, Voigt et al. (1996) reported the natural precipitation of hornesite ($Mg_3(AsO_4)_2 \cdot 8H_2O$), and Juillot et al. (1999) showed the occurrence of calcium arsenates and, in minor amounts, calcium-magnesium arsenates. Foster et al. (1998) found the formation of scorodite ($FeAsO_4.2H_2O$) in various mine wastes, but they also suggested that arsenates could be substituted for sulfates in the structure of jarosite ($KFe_3(SO_4)_2(OH)_6$). They also found the arsenate sorption on ferric oxyhydroxides and aluminosilicates. Davis et al. (1996) observed the formation of metal As oxides, FeAs oxides, and As phosphates in smelter-impacted soils and suggested that these As-bearing phases in Anaconda soils probably resulted from process wastes generated during historical smelting of copper sulfide ore in Anaconda.

As a general rule, we could consider that these secondary minerals are mainly composed of arsenates. Thus, Donahue et al. (2000) mention that 17,000 tons of As were discharged to the tailings management facility (Rabbit Lake uranium mine), of which approximately 15,000 tons (88%) were in the arsenate form and 2000 tons (12%) were in the form of primary arsenides. Stability of these secondary As-bearing phases is a function of several parameters. Three classes of arsenates may be distinguished: iron arsenates (usually encountered in the environment), calcium arsenates, and metal arsenates. Proportions of different arsenates are particularly a function of mine site activity (ores exploited, treatments used, etc.).

- In the presence of ferric iron in solution, ferric arsenates may be formed from As acid and precipitate (Reaction 10.2) (Lawrence and Marchant, 1988):

$$Fe_2(SO_4)_3 + 2H_3AsO_4 \rightarrow 2FeAsO_4 + 3H_2SO_4 \qquad (10.2)$$

Papassiopi et al., (1998) found that the optimum precipitation pH for achieving the lowest residual As in solution depends on the Fe/As ratio and precipitation temperature. At Fe/As = 4 and a precipitation temperature of 33°C, the optimum precipitation pH is 5, and the residual As in solution is less than 0.2 mg l⁻¹. They concluded that by increasing the Fe/As ratio, the optimum precipitation pH shifts to higher values, while by increasing the precipitation temperature, it shifts to lower ones. Krause et al. (1989) found that iron arsenates with molar ratios greater than four appear to have adequate stability. The ferrous compounds have lower solubilities in water than the corresponding ferric arsenates (Escobar Gonzales and Monhemius, 1988).

Of the ferric arsenates, only one anhydrous member, angelellite ($Fe(AsO_4)_2O_3$) has been reported. The other iron arsenate minerals are hydrated with different degrees of hydration. Scorodite ($FeAsO_4 \cdot 2H_2O$) is by far the most abundant and important of the iron arsenates. Scorodite is the result of oxidation of arsenopyrite (Reaction 10.3), loellingite, and realgar.

$$FeAsS + 14\ Fe^{3+} + 10H_2O \rightarrow 14Fe^{2+} + SO_4^{2-} + FeAsO_4 \cdot 2H_2O + 16H^+ \quad (10.3)$$

Studies of the solubilities of crystalline or amorphous iron minerals are not numerous (Escobar Gonzales and Monhemius, 1988). To date, the solubility of scorodite is the only one that has been measured, in both its amorphous and crystalline states and under different conditions (Makhmetov et al., 1981; Dove and Rimstidt, 1985). Krause and Ettel (1989) have shown that crystalline scorodite is more insoluble than amorphous scorodite. They also determined the solubility product, K_{sp}, of amorphous scorodite to be 3.89×10^{-25} (mol² l⁻²) for the pH range 0.97 to 2.43 ($pK_{sp} = 24.41 \pm 0.15$) and that, under comparable conditions, K_{sp} for crystalline scorodite is ~ 1000 times lower than the published value for amorphous $FeAsO_4 \cdot xH_2O$. Iron arsenates were used for immobilization of As in contaminated media (Artiola et al., 1990; Voigt et al., 1996).

•   In nature, there are calcium arsenate minerals, none of which is anhydrous except weilite ($CaHAsO_4$), which is an acid calcium arsenate anhydrate. Calcium arsenates currently encountered in nature are: pharmacolite $CaHAsO_4 \cdot 2H_2O$; haidingerite $CaHAsO_4 \cdot H_2O$. They could be formed by "natural" processes in an industrial area: acid waters interacting with the limestone substratum, providing dissolved calcium, which reacts with As to precipitate 1:1 arsenates and, in minor amounts, Ca-Mg arsenates (Juillot et al., 1999) These minerals could also be the result of treatment processes used to precipitate As (lime addition) following Reaction 10.4 (Collins et al., 1988):

$$H_3AsO_4 + Ca(OH)_2 \rightarrow CaHAsO_4 \cdot 2H_2O \quad (10.4)$$

According to paragenetic studies most of the calcium and calcium-magnesium arsenates were formed, and are stable, between pH 6 to 8, although some minerals, such as weilite, were formed at pH 3 to 5 (Escobar Gonzales and Monhemius, 1988). Most of the calcium arsenates are unstable in aqueous environments and are difficult to find in the upper parts of oxidation zones. Like iron arsenates, calcium arsenates are more insoluble than calcium arsenite for the same Ca/As ratio. To comply with environmental regulations a Ca/As ratio of 7 is required to achieve As solubility of 0.4 mg l⁻¹ (Stefanakis and Kontopoulos, 1988).

- The most wide-ranging investigation of the solubility of metal arsenates was by Chukhlantsev (1956), who reported the solubilities of 17 metal arsenates (and 6 metal arsenites; Chukhlantsev, 1957). Chukhlantsev showed that the solubility of metal arsenates is considerably lower than that of the corresponding arsenites.

A CASE STUDY

Experimental Procedure

Soil Collection and Preparation

The soil sample was collected from the surface soil (0 to 40 cm) of a former gold mine. A large mass of soil was required in this study, so 200 kg were sampled. Upon arriving in the laboratory, the soil was homogenized, air-dried, sieved to 2 mm, and stored at room temperature in polypropylene flasks until use. The soil was characterized with respect to its physicochemical properties, mineralogical composition, and As speciation.

Chemical Analysis

Deionized water (Millipore) was used throughout. All reagents were of analytical reagent grade or higher purity.

- Trace and major elements: Elements were solubilized using specific methods (total digestion by fluorhydric and perchloric acids for trace elements and alkaline digestion for major elements). These extracts were analyzed by an inductively coupled plasma-atomic emission spectrometer (ICP-AES).
- For total As determination, 0.5-g soil samples were dry-ashed at 450°C (2 h) with NH_4NO_3 (10%) and then, soils were dissolved in 35 ml of HCl (6 N) and were heated for 20 min. Solutions were diluted after filtration to 50 ml. Extracts were analyzed by ICP-AES.
- Arsenic speciation: The determination of four As species (As(III), dimethylarsinic acid (DMA), monomethylarsonic acid (MMA), and As(V)) was investigated using an on-line system involving high-performance liquid chromatography (HPLC) coupled with inductively coupled plasma mass spectrometry (ICP-MS) (Thomas et al., 1997). Phosphoric acid (1 M) was used in conjunction with an open focused microwave system to extract As compounds. For chromatographic separations, an anion exchange column was used.

XRD and SEM Analysis

X-Ray powder diffraction (XRD) was used to identify crystalline phases in the soil. Analyses were performed with a Philips PW1710 diffractometer using CuKα radiation ($\lambda = 1.5418$Å). XRD data were collected between 3 to 60° θ. Measurements were made using a step technique with a fixed time of 0.5 s by step. The fine fraction (< 2 μm) was observed with an Environmental Scanning Electron Microscope

(ENSEM model 2020) to analyze soil grains and matrix. Chemical composition data were acquired with an Energy Dispersive Spectrometer (EDS) (Oxford Instrument Link ISIS). Operating conditions for quantitative compositional analyses were 20 kV accelerating voltage and 2 T pressure in the sample chamber.

Sequential Extractions

A sequential extraction scheme was established according to modified Tessier et al. (1979) and Shuman (1985) schemes (Table 10.1). Such a scheme was used for As-contaminated soils by Gleyzes (1999). The experiments were conducted in centrifuge tubes with 2 g of soil. A continuous agitation was maintained during the extraction time. Between each successive extraction, separation was done by centrifugation (30 min at 4000 rd min^{-1}). The supernatant was removed and the residue was washed with 16 ml of deionized water. The extraction and wash supernatants were pooled. Arsenic was determined by graphite furnace atomic absorption spectrometry; Fe, Al, and Mn by flame atomic absorption spectrometry. The extractions were performed in triplicate.

Mobility Tests

Batch Experiments

- Arsenic remobilization batch experiments without pH control were conducted at room temperature in centrifuge tubes with 5 g of soil and 50 ml of solution. The As remobilization was studied over a range of pH values and phosphate concentrations that should be encountered in environmental conditions and all the extractions were done using the same electrolyte (1.76×10^{-4} M of $Ca(NO_3)_2$).

  The soil suspensions were shaken for the desired time with a rotary extractor. The resulting mixtures were centrifuged 30 min at 5000 rd min^{-1}. The supernatant was analyzed after 0.45 µm filtration for pH, As concentration, and other important parameters, depending on the experiments (Fe, Al, and Mn concentrations).

  The pH values were adjusted to 3, 5, 7, 9, 11, and 13 with 2 M or 5 M NaOH and 2 M HNO$_3$ solutions. The different phosphate concentrations (10^{-5}; 5.10^{-3}, and 2.10^{-2} M) were prepared with a 2 M K$_2$HPO$_4$ solution.

- For kinetics studies (batch tests with pH control), two independent experiments (at pH 3 and pH 11) were done with 200 g of soil and 2 liters of electrolyte solution. pH control was maintained by an automatic pH-titrimeter. At the desired time, an aliquot of supernatant (10 ml) was analyzed after 0.45 µm filtration for As, Fe, Mn, and Al concentrations.

Column Experiments

Two column tests were conducted to study As remobilization at pH 3 and pH 11. Saturated column experiments were operated in a PEEK (PolyEtherEtherKetone) column (1 cm internal diameter by 10 cm length) using an upflow mode. The small column size was chosen in order to perform the experiments in a reasonable time under controlled conditions. The column was packed with a determined amount of

Table 10.1   Sequential Extraction Procedure

Fractions	Reactants	Experimental Conditions
F1: Soluble in $MgCl_2$	$MgCl_2$ (1 mol l^{-1})	pH = 7, (8 ml, stirring 1 h)
F2: Bound to carbonates	CH_3COONa (1 mol l^{-1})/CH_3COOH (1 mol l^{-1})	pH = 4.5 (8 ml, stirring 15 h)
F3-Mn: Bound to Mn-oxides	$NH_2OH \cdot HCl$ (0.04 mol l^{-1}) in CH_3COOH 25%	pH = 2 (20 ml, 5 h 30 min (96°C))
F3-Fe,a: Bound to amorphous Fe oxides	$(NH_4)_2C_2O_4 \cdot H_2O$ (0.2 mol l^{-1})/$H_2C_2O_4$ (0.2 mol l^{-1})	pH = 2 (50 ml, 4 h stirring in darkness)
F3-Fe,c: Bound to crystalline Fe oxides	$(NH_4)_2C_2O_4 \cdot H_2O$ (0.2 mol l^{-1})/$H_2C_2O_4$ (0.2 mol l^{-1})/$C_6H_8O_6$ (0.1 mol.l^{-1}),	pH = 2 (50 ml, (100°C), 30 min)
F4: Bound to organic matter and sulfides	HNO_3 (0.02 mol l^{-1})/H_2O_2 30%	pH = 2 (3 ml HNO_3 + 2 ml H_2O_2: (85°C), 2 h then 2 ml H_2O_2: (85°C), 3 h)
	CH_3COONH_4 (3.2 mol l^{-1}) in HNO_3 20%	5ml CH_3COONH_4 in 20 ml of water, 30 min
F5: Residual	$HF/HClO_4$	10 ml HF/2 ml $HClO_4$ 10 ml HF/1 ml $HClO_4$ 1 ml $HClO_4$

Table 10.2   Soil Physical Chemical Characteristics

Soil Characteristic	Contaminated Soil
Moisture content *(%)	4.10
pH_{H2O}	6.80
TOC (%)	8.40
IC (%)	0.03
Clay (%)	3.80
Silt (%)	18.20
Sand (%)	78.00
Fe (%)	10.00
Al (%)	10.30
Mn (%)	0.09
Ca (%)	1.70
As (mg kg^{-1})	9400
Sb (mg kg^{-1})	270
Se (mg kg^{-1})	48
Hg (mg kg^{-1})	3.60
Zn (mg kg^{-1})	268
Cu (mg kg^{-1})	93
Pb (mg kg^{-1})	82
Cr (mg kg^{-1})	65
Cd (mg kg^{-1})	1.20
Ni (mg kg^{-1})	102

* residual moisture

dried soil. Leachates were collected by a fraction collector and an aliquot of each sample was analyzed for pH level and for total As, Fe, Mn, and Al contents by ICP-AES. The soil was saturated by injecting electrolyte solution ($1.76 \times 10^{-4} M$ of $Ca(NO_3)_2$) at the bottom of the column at a steady flow of 0.2 ml min^{-1} (HPLC pump). After 48 hours' equilibration, characteristic parameters of the column for each experiment were determined by pulse experiment using NO_3^- as a conservative tracer. Then the electrolyte was adjusted to the appropriated pH, and As remobilization was followed in the different fractions. Experiments were stopped when pH at outflow and inflow of the column were the same.

For the alkaline experiment, the pump, the column, and the fraction collector were put under inert atmosphere (N_2).

Results and Discussion

Soil Characterization

General soil characteristics and chemical analysis results are listed in Table 10.2. The mine soil with a pH of 6.8 has nearly no inorganic carbon, and it has a silt-sandy texture. The surface soil has a total As content of 9400 ± 350 mg kg^{-1} dry soil. The geochemical background for As in this area is around 200 mg kg^{-1}. Other heavy metals are at low concentrations. In addition, As was determined in five different particle size fractions (<2μm; 2 to 50 μm; 50 to 200 μm; 200 to 500 μm,

Table 10.3   Arsenic Speciation
in the Soil (%)

As Species	Abundance (%)
As(III)	9.3
As(V)	90.7
MMA	nd*
DMA	nd

* nd: nondetected

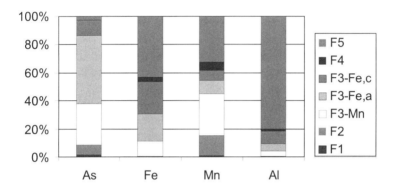

Figure 10.4   As, Fe, Mn, and Al released by sequential extractions (%).

and 500 to 2000 µm). The largest As percentage (35.2%) was found in the 50 to 200 µm fraction due to its large proportion in the soil.

Arsenic Speciation and Arsenic-Bearing Phases Determination

Analytical Chemical Speciation (HPLC-ICP-MS)

As chemical speciation was performed by HPLC-ICP-MS. The results (Table 10.3) show that both As(V) and As(III) occur in the soil and that As(V) represents 90.7% of total As. Furthermore, no methylated species were detected. These results confirm that As(V) compounds, more mobile and less toxic than As(III) compounds, predominate in aerobic soils.

Localization Phase Speciation (Sequential Extractions)

The first step of the extraction procedure ($MgCl_2$) is not able to easily replace adsorbed anions such as As(V) oxyanion, because Cl is only weakly adsorbed by soils. Arsenic distribution in the different fractions (Figure 10.4) shows that it seems to be easily mobilized. The sum of As percentages extracted in the first four fractions represents 86.5% of the total As. The fractions associated with the oxides contain 68.6% of the total As, with a large predominance of As in the fraction linked to the amorphous iron oxides (48.4%); then, in the fraction linked to the manganese oxides

Table 10.4	Particle (P) and Matrix (M) Compositions (% Atomic)	
Element	Atomic % in Particle (P)	Atomic % in Matrix (M)
As	7.6	0.4
Fe	5.9	3.4
O	66.7	58.4
Al	3.4	6.3
Si	3.6	9.6

(29.3%) and with only minor amounts in the crystallized iron oxides (10.2%). Most of the Fe was found in the noncrystalline and crystalline Fe oxide fractions as expected (44%), but a large amount was also found in the residual fraction (40%). Manganese is mostly solubilized in the residual fraction (33.5%) and in hydroxylamine hydrochloride (30.4%). Large amounts of metals are often found in the residual fraction in historically contaminated soils. (Voigt et al., 1996; Gleyzes, 1999).

As expected, almost all the Al is found in the residual fraction. We can therefore deduce that As is essentially associated with the amorphous iron oxides and to a lesser extent with the manganese oxides. We certainly are in the presence of an association of As/Fe. This association of As with oxides in contaminated soils has already been shown (Voigt et al., 1996; Dudas, 1987; Roussel et al., 2000) and, more precisely, the association of As/amorphous iron oxides (Gleyzes, 1999; McLaren et al., 1998).

Physical speciation (XRD-SEM)

The main components of the soil determined by X-ray diffraction were: quartz, feldspar, micas, and hematite in decreasing order of abundance. A crystallized iron oxide (hematite) was also found in the soil. Under the conditions used, no crystalline As mineral phase was detected by this technique (Foster et al., 1998; Voigt et al., 1996). We can, however, notice the presence of crystallized iron oxides (hematite). Sequential extractions and XRD measurements show that the As-bearing solid phases are noncrystalline and, therefore, are secondary mineral phases. In the studied fraction (<2 μm) an association of As and iron was found in particles and in the matrix soil by SEM observations (Figure 10.5). These two phases are mainly composed of O, As, Fe, Al, and Si (Table 10.4).

In the soil matrix, As was always associated with the aluminosilicate matrix and with iron. Arsenic was also present in minor amounts on agglomerates of particles or in individual particles. In this case, it is always associated with iron. In the 2 to 50 μm fraction, As was found to be primarily associated with the particles and not with the soil matrix. Investigations with SEM show an association between As and iron. These observations, however, don't allow us to distinguish between iron arsenates or As associated with Fe oxides (sorption or coprecipitation).

If iron arsenate is present, its stability depends both on its crystalline state and on the molar ratio of Fe/As. In the matrix, we observe a high Fe/As ratio, about 8.5,

A

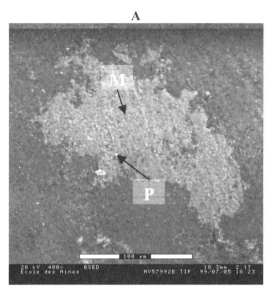

B

C

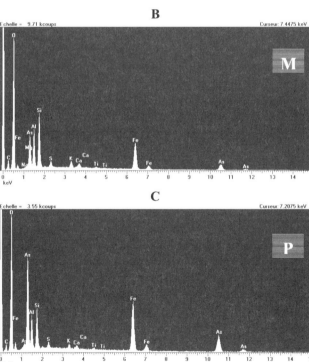

Figure 10.5 (A) Scanning electron microscopy (SEM) (backscattered-electron) of fine fraction of the contaminated soil. (B) EDS spectra of the matrix (M). (C) EDS spectra of the particle (P).

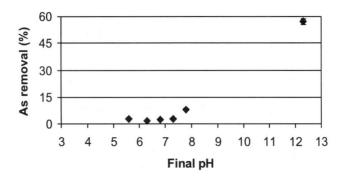

Figure 10.6   Arsenic remobilization in the supernatant as a function of pH (without pH control) (Stirring time = 24 h).

whereas the particle Fe/As ratio is about 0.8. Arsenic in these particles might be relatively mobilized. If we consider — as suggested by the whole characterization of As-bearing phases — that As seems to be mainly associated with amorphous mineral phases essentially composed of iron, we can infer that the mineral "neo-formed" species responsible for trapping As, in that soil are weakly stable.

Mobility Tests

pH Influence on Arsenic Remobilization

Arsenic remobilization was studied in batch (without and with pH control) and column experiments. Batch treatment is easier, faster, and more reproducible than column experiments. However, most batch experiments designed to examine a thermodynamic equilibrium do not consider the kinetics of metal/metalloid reactions with soil and cannot reflect hydraulic parameters such as preferential flow. So far, complexation properties of soils have only been discussed from a thermodynamic perspective. Equilibrium may never be reached under field conditions because of the rate of infiltration, which influences the contact time between the interstitial water and the solid phase (Van Der Sloot, et al., 1997). Arsenic retention depends on the reaction kinetics and the time of contact between As anions and soil. Columns may provide information that is not obtainable from batch experiments and is a more realistic simulation of field conditions.

Arsenic Remobilization as a Function of pH: Semiequilibrium Leaching Tests (batch tests without pH control)

Arsenic remobilization in the supernatant as a function of pH (Figure 10.6) shows an important soil buffer effect under acid conditions (initial pH going from 3 to 13). This experiment shows what happens when a sudden variation of an As-contaminated soil pH occurs. These tests allow us to assess risk when there is an accidental addition of acid or base.

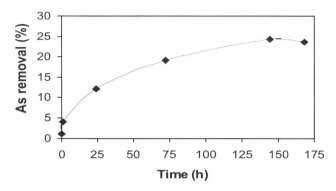

Figure 10.7   Kinetic of arsenic remobilization at pH 11 (with pH control).

Arsenic remobilization was higher under alkaline conditions (up to 57%). This observation matches the literature (Pantsar-Kallio and Manninen, 1997). In fact, if we refer to thermodynamic data, we notice that arsenates (As(V)) are in anionic forms from pH 2.4, whereas arsenites (As(III)) are in anionic forms above pH 9.2. (This As state is typically a minor species compared to arsenates.)

Under alkaline conditions, OH⁻ ions compete with As oxyanions for adsorption sites. Under acid conditions, As mobilization can be explained by the solubilization of iron arsenates (Krause, 1988). The solubility of basic iron arsenates $FeAsO_4 \cdot xH_2O$ is at a minimum for pH close to 4 (solubility on the order of 30 mg l⁻¹ of As); and as pH decreases, the solubility increases to 35 mg l⁻¹ at pH 3 and 500 mg l⁻¹ at pH 2. A solubilization of arsenates associated with iron can be inferred, but very little iron is solubilized in acid media, suggesting a rapid reprecipitation of As. Dove and Rimstidt (1985) suggest, for example, a dissolution of scorodite (common mineral of the arsenopyrite alteration often found in the surface environment) to form iron oxides (goethite) at a pH higher than 2 according to Reaction 10.5:

$$FeAsO_4 \cdot H_2O \rightarrow FeOOH + H_2AsO_4^- + H^+ \qquad (10.5)$$

Arsenic Remobilization as a Function of Time: Semiequilibrium Leaching Tests (batch tests with pH control)

Alkaline pH — An experiment was conducted to assess the kinetics of remobilization when the soil is in contact with an alkaline solution (pH 11). Arsenic was easily mobilized at pH 11 (Figure 10.7). Furthermore, equilibrium may have been reached after the 140th hour and the fraction remobilized increased with time and reached about 24% of total As after 140 h. Arsenic remobilization seemed to be stabilized after 144 h. Iron, aluminum, and manganese solubilization represents very low percentages (respectively at t = 144 h: 0.2%; 0.2% and 0.35%). These data match the thermodynamic data and the physicochemical equilibrium of these elements particularly for the iron oxides (Dzombak and Morel, 1990).

The use of a kinetics model in the study of sorption and desorption processes in heterogeneous systems continues to attract considerable interest. Three reasons

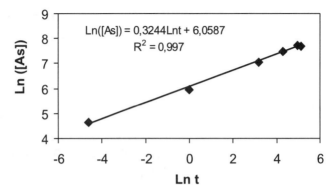

Figure 10.8 Kinetic study of arsenic: power function.

for the use of kinetic or time-dependent models in soils have been suggested (Skopp, 1986). First, many reactions in soils are slow, yet proceed at measurable rates. Slow reactions may be of great importance with regard to plant uptake and precipitation products. Second, nonequilibrium conditions can exist as a result of the physical transport of gases and solutes. Third, information about reaction mechanisms and processes occurring may be obtained from such data. Studies of the kinetics of pH reactions with As-contaminated soils are of interest for the mobility of As. Evaluation of kinetics models for As desorption was already studied in an artificially contaminated soil (Wasay et al., 2000). In our case, in a historically polluted soil, the remobilization of As over time was well described by a power function equation (Figure 10.8). The power function equation constants, b and k_d, were evaluated from the slope and intercept of the linear plots and found to be 0.3244 and 427.8, respectively, with an r^2 value of 0.997. The use of this function allows us to describe [As] removal (mg kg^{-1}) by the power function model (t in hour):

$$[As] = K_d \times t^b = 427.8 \times t^{\,0.3244}$$

Thus, in these conditions, arsenic is relatively mobile, and this "secondary source" of pollution presents a real risk for the environment.

Acid pH — Arsenic remobilization in acid medium (Figure 10.9), is much more important than expected, as shown in Figure 10.6. At 140 h, 21% of total As was solubilized. This percentage is similar to the one obtained at alkaline pH under the same conditions. Mobilization steadily increased and was linear ($r^2 = 0.9982$) in the studied time period (up to 140 h). Pantsar-Kallio and Manninen (1997) also showed an As remobilization under very acid conditions (pH 1) (and showed that at this extreme pH, As(III) was oxidized in As(V)). If we refer to arsenite and arsenate adsorption on iron oxides (Dzombak and Morel, 1990), we notice that at acid pH (pH < 3) only arsenites can be solubilized (maximum pH of adsorption from 3 to 7). We can conclude that the available stock of arsenite is remobilized.

In our case, As solubilization is accompanied by an important manganese solubilization (it can reach 42% of total manganese at t = 144 h), which seems to

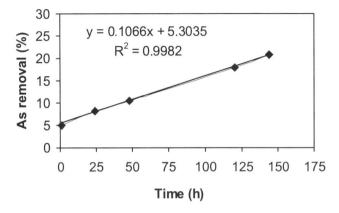

Figure 10.9 Kinetic of arsenic remobilization at pH 3 (with pH control).

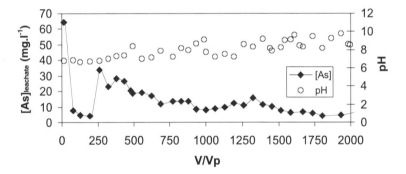

Figure 10.10 Leachability of total arsenic (mg l⁻¹) and pH evolution in a column test as a function of V/Vp under alkaline conditions (pH 11 injection at V/Vp = 300).

correspond to a manganese oxide solubilization. This element does not play a determining role, however, in the remobilization of As if we consider its low concentration in the studied soil (640 mg kg⁻¹). Iron and aluminum in solution were 3.6 and 2.3%, respectively, of the total. These values were higher than those measured under alkaline conditions, and the concentrations of these elements in solution are far from being insignificant if we consider the high percentages of iron and aluminum in the soil, respectively, 10.0% and 10.3% (cf. Table 10.2). In that case, we can infer a massive destruction of soil mineral phases and consequently a mobilization of As associated to these phases.

Dynamic Leaching Tests (Column)

Alkaline pH — Column tests have the advantage of being conducted under conditions approximating those observed in the field. Figure 10.10 represents the leachability of total As (mg l⁻¹) under alkaline conditions (electrolyte pH = pH 11) and pH evolution in a column test as a function of V/Vp. The average residence time in

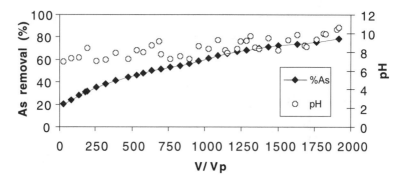

Figure 10.11   Cumulative curve of leachability of total arsenic (% of total arsenic) and pH
evolution in a column test as a function V/Vp under alkaline conditions (pH 11
injection at V/Vp = 300).

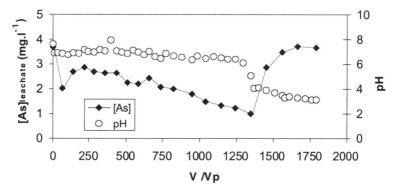

Figure 10.12   Leachability of total arsenic (mg l⁻¹) and pH evolution in a column test as a
function V/Vp under acid conditions (pH 3 injection at V/Vp = 213).

the column of the tracer used (NO_3^-) is about 33.5 min (Vp = 2.40) and electrolyte
injection at pH 11 occurred to V/Vp = 300 (after an equilibrium time of 48 h).

Just after the injection, we noticed an increase in As concentration at the column
outflow (Figure 10.10). The increase in As mobilization remained steady with time
(Figure 10.11).

At the end of this experiment, 72% of total As was remobilized. Since, As
remobilization steadily increased until the end of the experiment, we can infer that
the total As can potentially be mobilized when the contaminated soil is subjected
to a steady alkaline stress.

Acid pH — The average residence time in the column of the tracer used (NO_3^-)
was about 34.8 min (Vp = 2.66 ml) and electrolyte injection at pH 3 occurred to
V/Vp = 213. Figure 10.12 represents the leachability of total As (mg l⁻¹) under acid
conditions (electrolyte pH = pH 3) and pH evolution in a column test as a function
of V/Vp. In this experiment, pH influence (more precisely that of proton [H⁺] supply)
is clearly shown. An increase in As remobilization was observed when the pH

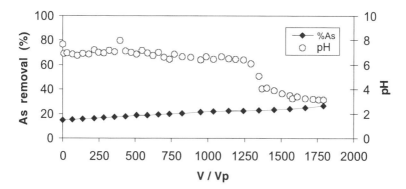

Figure 10.13   Cumulative curve of leachability of total arsenic (% of total arsenic) and pH evolution in a column test as a function V/Vp under acid conditions (pH 3 injection at V/Vp = 213).

suddenly decreased from 6.5 to 3. At that point, no additional H⁺ was "consumed" by the soil and the sites sorbing H⁺ were saturated. For a leaching volume similar to the one used in the alkaline experiment, we observed that 28% of As was remobilized (Figure 10.13) at the end of this experiment, compared to 72% under alkaline conditions. Thus, remobilization tests clearly show that As is very mobile under alkaline conditions. This agrees with other findings in the literature (Pantsar-Kallio and Manninen, 1997). In our case, As percentages extracted can reach up to 72% of total As in the column experiment. Whatever the pH conditions (alkaline or acid), a very fast and steady increase of As mobilization with time was observed.

Arsenic Remobilization as a Function of Phosphate Concentrations

As phosphates can be competitor anions of As for adsorption sites in soils, we have also done As remobilization tests in the presence of phosphates (batch and column experiments).

Semiequilibrium Leaching Tests (batch tests) — The influence of phosphate concentration in solution on As remobilization is illustrated in Figure 10.14 where As desorption increased with increasing solution concentration of P. The percentage of remobilized As reaches 6% in 24 h for a phosphate initial concentration of 2.10^{-2} M. Since addition of phosphate changes leachate pH, we have to take the final pH of supernatants into account to show a combined effect (phosphate concentration and pH) on As solubilization. The final pH measured during this experiment was 7.2. The process implied during As remobilization seems to be anion exchange. (Manful et al., 1989; Thanabalasingam and Pickering, 1986; Bhumbla and Keefer, 1994).

Dynamic Leaching Tests (column) — Figure 10.15 shows the leachability of total As (mg l⁻¹) at phosphate concentration of 2.10^{-2} mol l⁻¹ and phosphate concentration evolution in a column as a function of V/Vp. The average residence time in the column of the tracer used (NO₃⁻) is about 34.0 min (Vp = 2.49) and phosphate injection occurred to V/Vp = 270 (after an equilibrium time of 48 h).

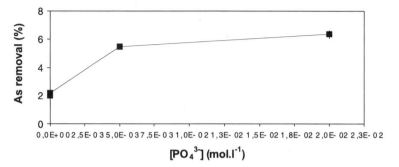

Figure 10.14   Arsenic remobilization in the supernatant as a function of phosphate concentrations (Stirring time = 24 h).

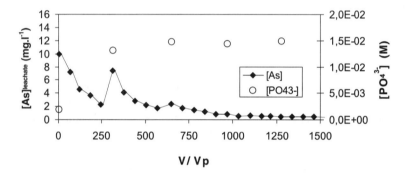

Figure 10.15   Leachability of total arsenic (mg l^{-1}) and [PO$_4^{3-}$] evolution in a column test as a function of V/Vp under phosphate conditions ([PO$_4^{3-}$] = 2.10^{-2} mol l^{-1} at V/Vp = 270).

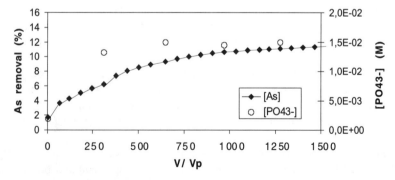

Figure 10.16   Cumulative curve of leachability of total arsenic (% of total arsenic) [PO$_4^{3-}$] evolution in a column test as a function of V/Vp under phosphate conditions ([PO$_4^{3-}$] = 2.10^{-2} mol l^{-1} at V/Vp = 270).

A sudden increase of As concentration at the column exit (up to 7 mg l⁻¹) was observed when phosphates were injected. The phosphate concentration tended toward 1.5 10^{-2} M, the As concentration decreased, to reach 0.35 mg l⁻¹ at the end of the experiment and 11.5% of total As was remobilized (Figure 10.16). Unlike the other column tests, As was strongly remobilized by the phosphate solution injection, which implies that anion exchange between As oxyanions and phosphate ions is an instant rapid reaction. The adsorption sites, at which the phosphate ions may exchange with As oxyanions, leads to a significant As remobilization of 11.5%. This site saturation seems to appear for a V/Vp of about 1000.

Finally, as previously noticed in the case of pH influence on As mobility, it appears that the dynamic tests (where the conditions are closer to the conditions on site) lead to As remobilization, which is more important than the semiequilibrium leaching tests (batch tests).

CONCLUSION

Arsenic remobilization in polluted soils is a function of several factors. It depends on As speciation in the soil: organic or inorganic forms; distribution of different species of As according to their oxidation state; nature of the arsenical minerals and the physicochemical conditions in the environment (pH, redox potential, occurrence of competitor anions of the As oxyanions, nature and abundance of mineral phases able to trap As, etc.). The purpose of this study was to clarify the roles of pH and phosphate ions in As remobilization in a historically polluted soil. The experiments on the determination of the As-bearing phases indicate that this element is essentially present in the pentavalent form (As(V)), associated with amorphous iron oxides. Sequential extractions predict that As is mobile at this site. This remobilization has been confirmed by different As remobilization tests (in batch with or without pH control and in column tests). A large amount of As remobilization under alkaline conditions (up to 72%) was observed. Under acid conditions, As solubilization, even though it was lower than under alkaline conditions, was significant (up to 28% of As in solution). Dynamic tests showed As remobilization from phosphate injection. As removal clearly decreased after the sites (which potentially could be occupied by phosphates) had been occupied.

The risks from As pollution should be seriously considered. Whatever the conditions studied and the type of remobilization experimentation carried out, the minimal concentrations of solubilized As are much higher than the maximum concentration level (MCL) for drinking water that has been established for As (50 μg l⁻¹).

ACKNOWLEDGMENTS:

The work reported in this paper was supported by the ADEME (French Agency for Environment and Energy Management) and the Water and Environment department of the Institut Pasteur de Lille. The authors gratefully acknowledge the CNRSSP (National Research Center for Polluted Sites and Soils).

REFERENCES

Agget J. and O'Brien G.A., 1985. Detailed model for mobility of arsenic in lacustrine sediments based on measurements in Lake Ohakuri, *Environ. Sci. Technol.*, 19, pp. 231-237.

Anderson M.A., Ferguson J.F. and Gavis J., 1975. Arsenate adsorption on amorphous aluminum hydroxide, *J. Colloid Interface Sci.*, 54, 3, pp. 391-399.

Artiola J.F., Zabcik D. and Johnson S.H., 1990. *In situ* treatment of arsenic contaminated soil from a hazardous industrial site: laboratory studies, *Waste Man.*, 10, pp. 73-78.

Azcue J.M., Mudroch A., Rosa F. and Hall G.E.M., 1994. Effects of abandoned gold mine tailings on the arsenic concentrations in water and sediments of Jake of Clubs Lakes, B.C., *Environ. Technol.*, 15, pp. 669-678.

Backer D.E. and Chesnin L., 1975. Chemical monitoring of soils for environment quality and animal and human health, *Adv. Agron.* 27, pp. 305-374.

Bhumbla D.K. and Keefer R.F., 1994. Arsenic mobilization and bioavailability in soils. In *Arsenic in the Environment. Part I: Cycling and Characterization*, (Ed. Nriagu J.O.), John Wiley & Sons, Inc., New York.

Brannon J.M. and Patrick W. H. Jr., 1987. Fixation, transformation, and mobilization of arsenic in sediments, *Envir. Sci. Technol.*, 21, pp. 450-459.

Braux C., Piantone P., Zeegers H., Bonnemaison M. and Prévot J.C., 1993. Le Châtelet gold-bearing arsenopyrite deposit, Massif Central, France: mineralogy and geochemistry applied to prospecting, *Appl. Geochem.*, 8, pp. 339-356.

Brookins D.G., 1988. *Eh-pH Diagrams for Geochemistry*, Springer-Verlag, New York, p. 175.

Cheng C.N. and Focht D.D., 1979. Production of arsine and methylarsines in soil and in culture, *Appl. Environ. Microbiol.*, 38, pp. 494-498.

Chukhlantsev V.G., 1956. The solubility product of metal arsenates, *Zhur. Neorg. Khim.*, 1, pp. 1975-1982.

Chukhlantsev V.G., 1957. The solubility of arsenious acid salts, *Zhur. Anal. Khim.*, 2, pp. 1190-1193.

Collins M.J., Berezowsky R.M. and Weir D.R., 1988. The behaviour and control of arsenic in the pressure oxidation of uranium and gold feedstocks. In *Arsenic Metallurgy, Fundamentals and Applications*. Eds. Reddy, Hendrix, Queneau. Ed. Metall. Soc. Inc.

Davis A., Houston Kempton J. and Nicholson A., 1994. Groundwater transport of arsenic and chromium at a historical tannery, Woburn, Massachusetts, U.S.A., *Appl. Geochem.*, 9, pp. 569-582.

Davis A., Ruby M.V., Bloom M., Schoof R., Freeman G. and Bergstrom P.D., 1996. Mineralogic constraints on the bioavailability of arsenic in smelter-impacted soils, *Environ. Sci. Technol.*, 30, pp. 392-399.

Deuel L.E. and Swoboda A.R., 1972. Arsenic solubility in a reduced environment, *Soil Sci. Am. Proc.*, 36, pp. 276-279.

Donahue R., Hendry M.J. and Landine P., 2000 Distribution of arsenic and nickel in uranium mill tailings, Rabbit Lake, Saskatchewan, Canada, *Appl. Geochem.*, 15, pp. 1097-1119.

Dove P.M. and Rimstidt J.D., 1985. The solubility and stability of scorodite, *Am. Min.*, 70, pp. 838-844.

Dudas M.J., 1987. Accumulation of native arsenic in acid sulfate soils in Alberta, *Can. J. Soil Sci.*, 67, pp. 317-331.

Dzombak D.A. and Morel F.M.M., 1990. *Surface Complexing Modeling. Hydrous Oxide*, John Wiley & Sons, Inc., New York, p. 393.

Escobar Gonzales V.L. and Monhemius A.J., 1988. The mineralogy of arsenates relating to arsenic impurity control. In *Arsenic Metallurgy, Fundamentals and Applications*. Eds Reddy, Hendrix, Queneau. Ed. Metall. Soc. Inc.

Ferguson J.F. and Gavis J., 1972. A review of arsenic cycle in natural waters, *Water Res.*, 6, pp. 1259-1274.

Foster A.L., Brown G.E. Jr., Tingle T.N. and Parks G.A., 1998. Quantitative arsenic speciation in mine tailings using X-ray absorption spectroscopy, *Am. Min.*, 83, pp. 553-568.

Frost R.R. and Griffin R.A., 1977. Effect of pH on adsorption of arsenic and selenium from landfill leachate by clays minerals, *Soil Sci. Soc. Am. J.*, 41, pp. 53-57.

Gleyzes C., 1999. Conditions de solubilisation et mise au point de schémas de caractérisation chimique de métaux et d'arsenic dans des sols de sites industriels et miniers, Ph.D. thesis, Thèse de doctorat de l'Université de Pau et des Pays de l'Adour, p. 236.

Goldberg S. and Glaubig R.A., 1988. Anion sorption on a calcareous, montmorillonitic soil-arsenic, *Soil Sci. Soc. Am. J.*, 2, pp. 1297-1300.

Gulens J., Champ D.R. and Jackson R.E., 1979. Influence of redox environments on the mobility of arsenic in ground water, *Am. Chem. Soc. Symp. Ser.*, pp. 81-95.

Gupta S.K. and Chen K.Y., 1978. Arsenic removal by adsorption, *Journal WPCF*, pp. 493-506.

Gustafsson J.P. and Tin N.T., 1994. Arsenic and selenium in some Vietnamese acid sulfate soils, *Sci. Total Environ.*, 151, pp. 153-158.

Harper M. and Haswell S.J., 1988. A comparison of copper, lead and arsenic extraction from polluted and unpolluted soils, *Environ. Technol. Lett.*, 9, pp. 1271-1280.

Juillot F., Ildefonse Ph., Morin G., Calas G., De Kersabiec A.M. and Benedetti M., 1999. Remobilization of arsenic from buried wastes at an industrial site: mineralogical and geochemical control, *Appl. Geochem.*, 14, pp. 1031-1048.

Krause E. and Ettel V.A., 1989. Solubilities and stabilities of ferric arsenate compounds, *Hydrometallurgy*, 22, pp. 311-337.

Lawrence R.W. and Marchant P.B., 1988. Biochemical pretreatment in arsenical gold ore processing, In *Arsenic Metallurgy, Fundamentals and Applications*. Eds. Reddy, Hendrix, Queneau. Ed. Metall. Soc. Inc.

Lin Z. and Puls R.W., 2000. Adsorption, desorption and oxidation of arsenic affected by clay minerals and aging process, *Environ. Geol.*, 39, 7, pp. 753-759.

Livesey N.T. and Huang P.M., 1981. Adsorption of arsenate by soils and its relation to selected chemical properties and anions, *Soil Sci.*, 131, 2, pp. 88-94.

Lund U. and Fobian A., 1991. Pollution of two soils by arsenic, chromium and copper, Denmark, *Geoderma*, 49, pp. 83-103.

Makhmetov M.Zh., Sagadieva A.K. and Chaprakov V.I., 1981. Solubility of iron arsenates, *J. Appl. Chem. USSR.*, 54(5), pp. 823-824.

Manful G.A., Verloo M. and De Spiegeleer F., 1989. Arsenate sorption by soils in relation to pH and selected anions, *Pedologie*, pp. 55-68.

Manning B.A. and Goldberg S., 1997a. Adsorption and stability of arsenic (III) at the clay mineral-water interface, *Environ. Sci. Technol.*, 31, pp. 2005-2011.

Manning B.A. and Goldberg S., 1997b. Arsenic(III) and arsenic(V) adsorption on three California soils, *Soil Sci.*, 162, 12, pp. 886-895.

Masscheleyn P.H., Delaune R.D. and Patrick W.H. Jr., 1991. Effect of redox potential and pH on arsenic speciation and solubility in a contaminated soil, *Environ. Sci. Technol.*, 25, pp. 1414-1419.

McBride M.B., 1994. *Environmental Chemistry of Soils*, Oxford University Press Inc., New York, p. 406.

McGeehan S.L. and Naylor D.V., 1994. Sorption and redox transformation of arsenite and arsenate in two flooded soils, *Soil Sci. Soc. Am. J.*, 58, pp. 337-342.

McLaren R.G., Naidu R., Smith J. and Tiller K.G., 1998. Fractionation and distribution of arsenic in soils contaminated by cattle dip, *J. Environ. Qual.*, 27, pp. 348-354.

O'Neill P., 1995. Arsenic, In *Heavy Metals in Soils*, (Ed. Alloway, B.J.), Blackie Academic & Professional, Glasgow.

Onken B.M.and Hossner L.R., 1996. Division S-2 — soil chemistry; determination of arsenic species in soil solution under flooded conditions, *Soil Sci. Soc. Am. J.*, 60, pp. 1385-1392.

Oscarson D.W., Huang P. M., Defosse C. and Herbillon A., 1981. Oxidative power of Mn(IV) and Fe(III) oxides with respect to As(III) in terrestrial and aquatic environments, *Nature*, 291, pp. 50-51.

Pantsar-Kallio M. and Manninen P.K.G., 1997. Speciation of mobile arsenic in soil samples as a function of pH, *Sci. Total Environ.*, 204, pp. 193-200.

Papassiopi N., Stefanakis M. and Kontopoulos A., 1988. Removal of arsenic from solutions by precipitation as ferric arsenates. In *Arsenic Metallurgy, Fundamentals and Applications*, Eds. Reddy, Hendrix, Queneau. Ed. Metall. Soc. Inc.

Piantone P., Wu X. and Touray J.C., 1994. Zoned hydrothermal alteration and genesis of the gold deposit at Le Châtelet (French Massif Central), *Econ. Geol.*, 89, pp. 757-777.

Pierce M.L.and Moore C.B., 1980. Adsorption of arsenite on amorphous iron hydroxide from dilute aqueous solution, *Environ. Sci. Technol.*, 14, 2, pp. 214-216.

Pierce M.L. and Moore C.B., 1982. Adsorption of arsenite and arsenate on amorphous iron hydroxide, *Water Res.*, 16, pp. 1247-1253.

Roussel C., Bril H. and Fernandez A., 2000. Arsenic speciation: involvement in evaluation of environmental impact caused by mine wastes, *J. Environ. Qual.*, 29, pp. 182-188.

Sadiq M., 1997. Arsenic chemistry in soils: an overview of thermodynamic predictions and field observations, *Water Air Soil Pollut.*, 93, pp. 117-136.

Sadiq M., Zaidi T.H. and Mian A.A., 1983. Environmental behaviour of arsenic in soils: theoretical, *Water Air Soil Pollut.*, 20, pp. 369-377.

Sadler R., Olszowy H., Shaw G., Biltoft R. and Connell D., 1994. Soil and waters contamination by arsenic from a tannery waste, *Water Air Soil Pollut.*, 78, pp. 189-198.

Sheppard S.C., 1992. Summary of phytotoxic levels of soil arsenic, *Water Air Soil Pollut.*, 64, pp. 539-550.

Shuman L.M., 1985. Fractionation method for soil microelements, *Soil Sci.*, 140, 1, pp. 11-22.

Skopp J., 1986. Analysis of time-dependent chemical processes in soils, *J. Environ. Qual.*, 15, 3, pp. 205-213.

Smith E., Naidu R. and Alston A.M., 1998. Arsenic in the soil environment: a review, *Adv. Agron.*, 64, pp. 149-195.

Stefanakis M. and Kontopoulos A., 1988. Production of environmentally acceptable arsenites-arsenates from solid arsenic trioxide. In *Arsenic Metallurgy, Fundamentals and Applications*. Eds. Reddy, Hendrix, Queneau. Ed. Metall. Soc. Inc.

Sun X. and Doner H.E., 1996. An investigation of arsenate and arsenite bonding structures on goethite by FTIR, *Soil Sci.*, 161, 12, pp. 865-872.

Tessier A., Campbell P.G.C. and Bisson M., 1979. Sequential extraction procedure for speciation of particulate trace metals, *Anal. Chem.*, 51, 7, pp. 844-851.

Thanabalasingam P. and Pickering W.F., 1986. Arsenic sorption by humic acids, *Environ. Pollut.*, 12, pp. 233-246.

Thomas P., Finnie J.K. and Williams J.G., 1997. Feasibility of identification and monitoring of arsenic species in soil and sediment samples by coupled high-performance liquid chromatography-inductively coupled plasma mass spectrometry, *J. Anal. At. Spectrom.*, 12, pp. 1367-1372.

Van Der Sloot H.A., Heasman L. and Quevauvillier Ph., 1997. *Harmonization of Leaching/Extraction Tests*, Studies in Environmental Science 70, Elsevier, p. 281.

Van Olphen H., 1963. An introduction to clay colloid chemistry, *Interscience*, New York., pp. 89-119.

Voigt D.E., Brantley S.L. and Hennet R.J.-C., 1996. Chemical fixation of arsenic in contaminated soils, *Appl. Geochem.*, 11, pp. 633-643.

Wakao N., Koyatsu H., Komai Y., Shimokawara H., Sakurai Y. and Shiota H., 1988. Microbial oxidation of arsenite and occurrence of arsenite-oxidizing bacteria in acid mine water from a sulfur-pyrite mine, *Geomicrobiol. J.*, 6, pp. 11-24.

Wasay S.A., Parker W., Van Geel P.J., Barrington S. and Tokunaga S., 2000. Arsenic pollution of a loam soil: retention form and decontamination, *J. Soil Contam.*, 9(1), pp. 51-64.

Xu H., Allard B. and Grimvall A., 1991. Effects of acidification and natural organic materials on the mobility of arsenic in the environment, *Water Air Soil Pollut.*, 57-58, pp. 269-278.

Xu H., Allard B. and Grimvall A., 1991. Influence of pH and organic substance on the adsorption of As(V) on geologic materials, *Water Air Soil Pollut.*, 40, pp. 293-305.

CHAPTER 11

Chemical Structures of Soil Organic Matter and Their Interactions with Heavy Metals

William L. Kingery, André J. Simpson, and Michael H.B. Hayes

INTRODUCTION

The organic components of soils and waters react with metal oxyhydroxide and clay minerals to form combinations of widely diverse chemical and biological stabilities (Schnitzer and Kodama, 1977). These associations, or organomineral complexes, profoundly affect the moisture and aeration regime, surface properties, biological activity, and many other environmentally important reactions occurring in these systems. The chemical composition and structure, as well as the surface properties of natural organomineral colloids are directly related to the mechanisms associated with their interactions with heavy metals in soils. Molecular structures, i.e., organic and inorganic functionalities, largely dictate the reactivities of heavy metals and their consequent transport with water percolating through the soil profile and/or their participation in chemical and biological reactions. Sorption and release, i.e., interactions, of heavy metals in the soil environment depend on their chemical forms and properties and their relative distribution in solution and solid phases (Förstner, 1991). Often, a large percentage of heavy metals are associated with soil solids (Hesterberg et al., 1993; Han et al., 2000). The control of heavy metal partitioning by soil colloids is key to determining their influence on contaminant bioavailability, toxicity, and persistence in the soil-water-sediment system (Di Toro et al., 1992).

The focus here is with natural organic matter because of the important role it plays in heavy metal chemistry in most soils. The complexity of the soil matrix presents an obstacle to our understanding of the basic processes involved in heavy metal reactivity in soils. Although it is not necessary to know molecular structures of organic matter to obtain an acceptable understanding of the gross interactions involving heavy metals and soil colloids, it will not be possible, nevertheless, to provide unambiguous reaction mechanisms until there is a better awareness of the component molecules, of the linkages between these molecules, and of the juxtaposition and spacings of reactive functional groups (Hayes and Swift, 1978). Recent developments in techniques, such as high-resolution NMR spectroscopy, have led to dramatic improvements in the description of natural soil organic matter components, making possible clearer interpretations of sorption and release of heavy metals by these compounds (Simpson et al., 1997a,b,c, 1999, 2001; Kingery et al., 2000).

REACTIVITY OF HUMIC SUBSTANCES AND HEAVY METALS

Metal-NMR

Chemical shifts of NMR-active metal ions (e.g., Cd, Al, and Pb) are sensitive to differences in their local environments and can therefore be of potential use in metal-binding studies to signify the degree of coordination, the various species and numbers of ligands, the functions of binding structures, and reaction kinetics and equilibria. Metal chemical shifts of complexes are affected by metal:humic substances molar ratios, solution pH, solvent properties, and lifetime of the complexes. The basis of metal-chemical shift changes is the interaction of the electron cloud surrounding each nucleus with the external magnetic field (Macomber, 1998). Electron clouds *shield* nuclei from the magnetic field so that shielded nuclei experience an effective field strength that is reduced relative to what would be experienced were the clouds not present. The greater the shielding, the lower (further upfield) will be the resonance frequency or chemical shift value, δ. When metal nuclei bind to humic structures, electrons are withdrawn, i.e., they are *deshielded*, by electronegative ligand moieties containing O, N, and S donor atoms (Senesi, 1992; Li et al., 1998). Based strictly on Pauling electronegativity values, the degree of deshielding would be arranged as $S(2.5) < N(3.0) < O(3.5)$. But, the specific structures wherein these atoms reside (e.g., carboxyl — $RCOO^-$; amine — R_3N; thiol — RS^-) alter the overall deshielding effect (Kostelnik and Bothner-By, 1974; Grassi et al., 1996). For example, the downfield shift, or deshielding, sequence for ^{113}Cd bound to donor atoms in a variety of proteins becomes O (δ ca. 100 to 150 ppm) < N (δ ca. 30 to 280 ppm) < S (δ ca. 750 ppm) (Summers, 1988).

Cadmium-113 NMR spectroscopy has been used to examine the coordination environment of Zn, Ca, Cu, Hg, Mn, and Mg binding sites in metalloproteins (Summers, 1988). The efficacy of ^{113}Cd as a metallobioprobe is related to a number of advantageous properties: 1) the ability to form a wide variety of conformational and multiligand complexes; 2) a nuclear spin number of ½ so that problems due to quadrupolar relaxation effects, i.e., line broadening, are avoided; 3) a relatively desirable magnetogyric ratio (γ) of -59.531×10^6 radians tesla^{-1} s^{-1} (compared to γ for ^{13}C of 67.264×10^6 radians tesla^{-1} s^{-1}); and, 4) a chemical shift range of over 900 ppm, the value of which has been shown to be sensitive to the nature, number, and geometric arrangement of the atoms coordinated to Cd (Summers, 1988). Cadmium-113 NMR offers a useful means for characterizing humic substances binding sites (Chung et al., 1996; Larive et al., 1996; Grassi and Gatti, 1998; Li et al., 1998).

The complexation of Cd to a Suwannee River fulvic acid was compared with complexes formed with EDTA because EDTA: 1) provides homogeneous Cd complexes; 2) has the capability of complexing all of the Cd present and thereby serves as a reference point for bound Cd; 3) negative and O-containing functional groups allow assessment of the binding information inherent in [113]Cd chemical shift data; and, 4) differences from the EDTA-[113]Cd spectra may give a relative sense of the presence or absence of binding by S-containing structures. The results revealed several facets of Cd-fulvic acid interactions at different [113]Cd:fulvic acid ratios. At a lower ratio a relative chemical shift of +35.6 ppm implies a fast chemical exchange of [113]Cd between binding sites and bulk solution (data not shown). Fast exchange is inferred from the lack of two distinct peaks representing both bound and unbound [113]Cd, and so the one observed is a weighted average of the resonance signals from the two populations of [113]Cd nuclei (Connors, 1987). And, for a higher Cd:fulvic acid ratio, where the species are mostly bound [113]Cd, the measured chemical shift relative to that of [113]Cd-EDTA (−66.3 ppm) suggests an environment with much less deshielding than the O-N coordinative array associated with Cd-EDTA complexes. The implication of these results is that O structures are the predominant binding functionalities in this sample.

Both Chung et al. (1996) and Li et al. (1998), however, produced [113]Cd-NMR evidence for the involvement of N-containing structures, as well as those containing O. Li et al. (1998) conducted experiments at basic pH values, which produced direct evidence of Cd binding by N donor atoms. They found for the same fulvic acid material mentioned above a broad peak at −10 to −30 ppm associated with the O-containing structures binding to Cd as well as an additional peak further downfield (between +40 and +70 ppm). The downfield peak was deemed to be Cd complexed by N structures since it was within the range of N deshielding and could not be assigned to unbound Cd species capable of existing under the conditions of their experiment.

Aggregation by Metals

Humic and fulvic acids are known to coagulate and precipitate from solution upon interaction with electrolytes (Stevenson, 1994). Based on the discussion above regarding associations of humics, it is postulated that cations are involved in aggregation of components. In addition, this process ought to be related to the structures in humic mixtures as well as chemical properties of the cation. Information on these processes would be highly relevant to the development of quantitative heavy metal release and mobility characteristics.

Humic samples were taken from A_h and B_h horizons of podzols at an Irish oak forest site and an adjacent site where the forest had been cleared and cultivated for 400 years. After sequential exhaustive extraction, the humic substances were isolated on XAD-8 and XAD-4 resins in tandem (Malcolm and MacCarthy, 1992; Hayes, 1996; Simpson, 1997b). Metal-nitrate standards were added to stock solutions of humic substances dissolved in distilled water. The solutions were then equilibrated overnight and filtered through 0.2-μm filters and the filtrates analyzed for metal content by ICP and for DOC by a C-analyzer.

In all cases, precipitation was observed with 0.008 M metal concentrations. The amount of fulvic precipitation was seen to occur in the order Cu > Al > Ca > Zn > Mg > K > Li. With the exception of Cu, this order is approximately trivalent > divalent > monovalent metal ion. Copper has a relatively high charge-to-radius ratio and is known to form strong bonds with humic substances (Stevenson, 1994). Further, Cu is able to complex with both O- and N-donor groups in humics, which may also help explain extensive precipitation of humic substances caused by Cu. Interestingly, while Cu produced considerable precipitation, only small amounts of the metal were removed from solution. On the other hand, Zn caused moderate precipitation but was readily removed from solution by the humic precipitate. This phenomenon is likely related to the hydrated nature of Zn. The solvation shell may act like a charged sphere, which will readily interact with humics, but may not form strong enough bridging to bring about the precipitation observed with Cu.

With Cu, the humic substances precipitated in the expected order humic acids > fulvic acids > XAD-4 acids (the latter are the materials retained on XAD-4 resins). The amount of complexed metal with fulvic acids is proportional to the amount of precipitate. The A_h horizon fulvic acid produced twice as much precipitate and bound twice as much Cu, when compared to that from the B_h horizon. The same result did not hold for the XAD-4 acids. This is due to the hydrophilic nature of the XAD-4 acids, which tend to remain in solution under all conditions. Additionally, the high proportion of phenolic components in XAD-4 acids may help explain their high affinity for Cu. The humic acids behaved differently in that the amount of precipitation was not directly related to the amount of Cu removed from solution. With both humic and fulvic acids the amount of Cu removed from solution in no way correlated with either CEC or quantities of functional groups within the materials (Simpson, 1999). These findings indicate that overall structure and conformation will need to be considered along with composition in order to predict and understand the mechanisms of their interactions with heavy metals. These findings highlight the potential problems in trying to obtain metal binding constants. In some cases, the precipitates formed at less than 5 ppm ion concentration and were difficult to see with the naked eye. The calculation of binding constants in a precipitating system is likely to lead to misleading results.

METAL CHELATE AFFINITY CHROMATOGRAPHY

Over the last few years, iminodiacetic acid supported on column resins has been used for the purification of proteins. The functionality can be loaded with any di- or trivalent metal which bridges across the carboxyl groups in the resin. Dissolved humic substances, e.g., from soil drainage waters, can then pass down the column and components of the mixtures be separated on the basis of their affinity for the resin-bound metal species. Water washes from oak forest, grassland, and pine forest soils were passed down columns loaded separately with uranium, lead, and mercury salts. The quantities of organic matter sorbed from each site and to each specific metal are summarized in Table 11.1. After additional washing with 10 column volumes of 0.01 M NaCl to remove any unbound organic matter residues, the columns were eluted

Table 11.1   The Percentages of Dissolved Organic
Matter from Various Sources Retained
on an Aminodiacetic Column Loaded
with Different Metals

	Pine Forest	Oak Forest	Grassland
Mercury	73%	71%	77%
Uranium	41%	35%	50%
Lead	25%	29%	26%

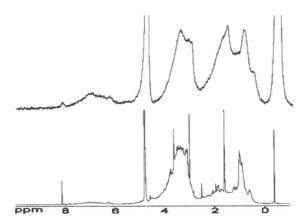

Figure 11.1 ^{1}H NMR spectra (A) the grassland organic matter that was retained on the column with uranium as the immobilized ion, and (B) the material that was not retained.

with 0.01 M Na-pyrophosphate. Initial ^{1}H-NMR the metal-complexed organic materials indicate very broad lines common to macromolecules or large aggregate structures (see Figure 11.1). Additional NMR experiments are needed in order to optimize conditions associated with the presence of complexed metals. Metal chelate affinity chromatography appears to be a widely applicable and versatile tool for the study of the complexation of metals by mobile organic components soil systems. Future studies are likely to employ the technique to help answer questions of metal complexation and mobility associated with soil mixtures.

FUTURE DIRECTIONS

Since soil is a key component controlling the behavior of trace metals in the environment, it is critically important to understand the influence of various soil constituents on fate and mobility. Humic substances, for example, are the principal organic components of soils and react with metal oxyhydroxides and clay minerals to form associations of widely diverse chemical and biological reactivities. The adsorption and release of heavy metals by natural organomineral complexes is complicated and is difficult to predict from consideration of the characteristics of individual components. It appears from the literature (not reviewed here) that the addition of sorption by isolated components provides a poor match to that determined for intact particles. These observations are consistent with a concept of organomineral structures whereby the functional groups, which are involved in metal binding as individual components, have altered reactivities as a result of the interactions leading to the formation of complexes.

With the advances made in our understanding of natural organic matter, it is now possible to construct models of these compounds with reasonable expectations of accuracy. These can be used to better represent natural soil and sediment particles in structure-reactivity studies of heavy metal behavior in the environment. Further, advances in analytical capabilities, for example high resolution magic angle spinning NMR, permit study of whole soil samples using a range of NMR experiments. In other words, multidimensional NMR experiments can be applied to naturally occurring samples, thereby eliminating concerns associated with obtaining representative samples using current methods of isolation and purification.

REFERENCES

Chung, K.H., S.W. Rhee, H.S. Shin, and C.H. Moon. 1996. Probe of cadmium(II) binding on soil fulvic acid investigated by [113]Cd NMR spectroscopy. *Can. J. Chem.* 74:1360-1365.

Connors, K.A. 1987. *Binding Constants: The Measurement of Molecular Complex Stability.* John Wiley & Sons, New York.

Di Toro, D.M., J.D. Mahony, D.J. Hansen, K.J. Scott, A.R. Carlson, and G.T. Ankley. 1992. Acid volatile sulfide predicts the acute toxicity of cadmium and nickel in sediments. *Environ. Sci. Technol.* 26:96-101.

Förstner, U. 1991. Soil pollution phenomena — mobility of heavy metals in contaminated soil. pp. 543-582. In G.H. Bolt, M.F. de Boodt, M.H.B. Hayes, and M.B. McBride (Eds.) *Interactions at the Soil Colloid-Soil Solution Interface,* Kluwer Academic Publishers: Dordrecht, Netherlands.

Grassi, M., E. Oldani, and G. Gatti. 1996. Metal-humus interaction: NMR study of model compounds. *Ann. Chim.* (Rome) 86:353-357.

Grassi, M. and G. Gatti. 1998.[113]Cd-NMR of model compounds for the study of metal-humus interaction. *Ann. Chim.* (Rome) 88:537-543.

Han, F.X., W.L. Kingery, H.M. Selim, and P.D. Gerard. 2000. Accumulation of heavy metals in a long-term poultry waste-amended soil. *Soil Sci.* 165:260-268.

Hayes, M.H.B. and R.S. Swift. 1978. The chemistry of soil organic colloids. p.179-320. D.J. Greenland, and M.H.B. Hayes (Eds.). In *The Chemistry of Soil Constituents.* John Wiley & Sons, Chichester.

Hayes, T.M. 1996. Isolation and characterisation of humic substances from soil, and the soil solution, and their interactions with anthropogenic organic chemicals. Ph.D. Thesis. The University of Birmingham.

Hesterberg, D., J. Bril, and P. del Castilho. 1993. Thermodynamic modeling of zinc, cadmium, and copper solubilities in a manured, acidic loamy-sand topsoil. *J. Environ. Quality.* 22:681-688.

Kingery, W.L., A. J. Simpson, M.H.B. Hayes, M.A. Locke, and R.P. Hicks. 2000. The application of multidimensional NMR to the study of soil humic substances. *Soil Sci.* 165:483-494.

Kostelnik, R.J., and A.A. Bothner-By. 1974. Cadmium-113 nuclear magnetic resonance studies of cadmium(II)-ligand binding in aqueous solutions. I. The effect of divers ligands
on the cadmium-113 chemical shift. *J. Magn. Reson.* 14:141-151.

Larive, C.K., A. Rogers, M. Morton, and W.R. Carper. 1996. [113]Cd NMR binding studies of
Cd-fulvic acid complexes: evidence of fast exchange. *Environ. Sci. Technol.*
30:2828-2831.

Li, J., M.E. Perdue, and L.T. Gelbaum. 1998. Using cadmium-113 NMR spectrometry to
study metal complexation by natural organic matter. *Environ. Sci. Tech.* 32:483-487.

Macomber, R.S. 1998. A complete introduction to modern NMR spectroscopy. John Wiley
& Sons, New York.

Malcolm, R.L. and P. MacCarthy. 1992. Quantitative evaluation of XAD-8 and XAD-4 resins
used in tandem for removing organic solutes from water. *Environment Internat.*
18:597-607.

Schnitzer, M., and H. Kodama. 1977. Reactions of minerals with soil humic substances.
p. 741-770. In J.B. Dixon and S.B. Weed (Eds.) *Minerals in Soil Environments.* Soil Sci.
Soc. Am., Madison, WI.

Senesi, N. 1992. Metal-humic substance complexes in the environment. Molecular and mechanistic aspects by a multiple spectroscopic approach. pp. 429-426. In D.C. Adriano (Ed.).
Biogeochemistry of Trace Elements. Lewis Publishers, Boca Raton, FL.

Simpson, A.J., R.E. Boersma, W.L. Kingery, R.P. Hicks, and M.H.B. Hayes. 1997a. Applications of NMR spectroscopy for studies of the molecular composition of humic substances. p. 46-62. M.H.B. Hayes, and M.A. Wilson (Ed.) In *Humic Substances, Peats
and Organic Amendments: Implications for Plant Growth and Sustainable Environment.*
The Royal Chemical Society, Cambridge, UK.

Simpson, A.J., B.E. Watt, C.L. Graham, and M.H.B. Hayes. 1997b. Humic substances from
podzols under oak forest and a cleared forest site I. Isolation and characterisation.
pp. 73-83. In M.H.B. Hayes and W.S. Wilson (Eds.) *Humic Substances, Peats and
Sludges: Health and Environmental Aspects.*, The Royal Society of Chemistry, Cambridge.

Simpson, A.J., T.M. Hayes, B.E. Watt, M.H.B. Hayes, and W.L. Kingery. 1997c. The application of advanced NMR spectroscopy for the identification of nitrogen and structure in
humic substances. In *Managing Risks of Nitrates to Humans and the Environment.*
Masterclass Conference organized by the Royal Society of Chemistry Agricultural Sector
and Toxicology Group, 1-2 September, University of Essex, Wivenhoe Park, Colchester.

Simpson A.J., 1999. The structural interpretation of humic substances isolated from podzols
under varying vegetation. Ph.D. Thesis, The University of Birmingham, England.

Simpson, A.J., J. Burdon, C.L. Graham, N. Spencer, M.H.B. Hayes, and W.L. Kingery. 2001.
Interpretation of heteronuclear and multidimensional NMR spectroscopy as applied to
humic substances. *Eur. J. Soil Sci.* (in press).

Stevenson, F.J. 1994. *Humus Chemistry: Genesis, Composition, Reactions. 2nd edition.* John
Wiley & Sons, New York.

Summers, M.F. 1988. [113]Cd NMR spectroscopy of coordination compounds and proteins.
Coordin. Chem. Rev. 86:43-134.

Index